TIME, SPACE AND PACE:
Computer-integrated Learning
in Corporate South Africa

This book is dedicated

To my children, the reason I live

To my mother, the reason I persevere

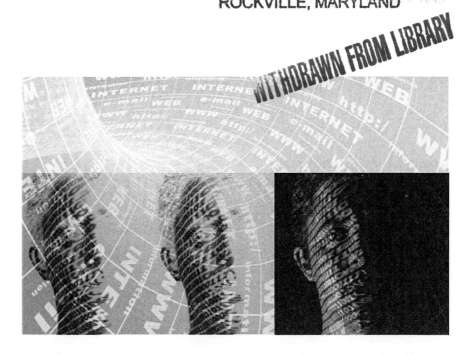

TIME, SPACE AND PACE:
Computer-integrated Learning
in Corporate South Africa

Rabelani Dagada

Unisa Press, Pretoria

ISBN 978-1-86888-459-9

Published by Unisa Press
University of South Africa
P O Box 392, 0003 UNISA

Book Designer: Elsabé Viljoen
Editor: Liz Stewart
Typesetting: Karen Graphics
Printer: Harry's Printers, Pretoria

CONTENTS

TABLES

FIGURES

ACRONYMS

ABET	Adult Basic Education and Training	HRD	Human Resources Development Strategy
ADDIE	Analysis, Design, Development, Implementation and Evaluation	HTML	Hypertext Markup Language
AfriNIC	African Network Information Centre	ICANN	Internet Corporation for Assignment of Names and Numbers
AsgiSA	Accelerated and Shared Growth Initiative for South Africa	ICT	Information and Communication Technology
BANKSETA	Banking Sector Education and Training Authority	IT	Information Technology
		Jipsa	Joint Initiative for Priority Skills Acquisition
BEE	Black Economic Empowerment		
BMR	Bureau of the Market Research (UNISA)	LAN	Local Area Network
BPO	Business Process Outsourcing	LCMS	Learning Content Management System
CCTLD	Country Code Top Level Domain	NALEDI	National Labour, Economic and Development Institute
CIA	Computer-Integrated Assessment		
CIE	Computer-Integrated Education	NCTU	National Council of Trade Unions
CIL	Computer-Integrated Learning	NGOs	Non-Governmental Organisations
COSATU	Congress of South African Trade Unions	NIA	National Intelligence Agency
		NQF	National Qualifications Framework
CRM	Customer Relationship Management	NSBs	National Standards Bodies
DBSA	Development Bank of Southern Africa	NUM	National Union of Mineworkers
DDBP	Design and Development Blueprint	OBET	Outcomes-Based Education and Training
DIDTETA	Defence and Trade Education Training Authority	RPL	Recognition of Prior Learning
		SACCAWU	South African Catering and Commercial Allied Workers Union
DoE	Department of Education		
DoHA	Department of Home Affairs	SAQA	South African Qualifications Authority
DoL	Department of Labour	SARS	South African Revenue Service
EAP	Employment Assistance Programme	SETAs	Sector Education and Training Authorities
ECT Act	Electronic Communications and Transactions Act		
		SMMEs	Small, Medium and Micro Enterprises
ETQAs	Education and Training Quality Assurers	UNISA	University of South Africa
FET	Further Education and Training	UWC	University of the Western Cape
GDE	Gauteng Department of Education	UWUSA	United Workers Union of South Africa
GEM	Global Entrepreneurship Monitoring	WAN	Wide Area Network

FOREWORD

By David Redshaw

It would be difficult, if not impossible, to image anyone better suited to write a book that focuses on computer-integrated learning in corporate South Africa than a young scholar who studied in eminent schools such as Wits University and the University of London. Rabelani Dagada has written extensively on Information Communication Technologies (ICT) and learning, as a glance at the Google Scholar shows. This book unpacks major challenges facing South Africa and suggests how computers can be employed to improve the lives of the people. It should be noted that corporate South Africa has acknowledged some of the assertions and observations that are made in this book and thus they are addressing the need to bridge the digital divide. Through the Community Social Investments, some organisations in South Africa are also closing the gap between the first and second economies and integrating the marginalised societies into the mainstream of ICT. But still, these well-intended efforts do not yet address the problem of skills shortage in this country sufficiently. The ripple effect of this book will require extensive reassessment of our interventions regarding the skills shortage.

Dagada was able to travel to urban and underdeveloped areas of South Africa and interacted with professionals in various economic sectors such as mining, banking, telecommunications, energy and services. He should be commended for not just relying on the available literature. The author matches up real life situations in corporate South Africa to the existing literature.

For some time Dagada has studied and focused on main challenges, broadening and detailing an outline of specific interventions that computer-integrated learning can adapt to better address the transformation of human resources development. It is interesting to note that he does not just espouse the benefits of computers, but also reflects on their weaknesses in a training and learning environment. He avoids singling out the use of computers for human resources development purposes as a panacea of all our skills shortages and poor productivity. It should also be noted that not all scholars support unconditionally the use of computers for teaching and learning. In an editorial of the journal *Education as Change (ICT in Education, 2005)*, Duan van der Westhuizen, who is a professor of educational ICT at the University of Johannesburg, noted that the notion that the use of computers 'leads to improved education has been under scrutiny for several years'. Van der Westhuizen supported his claim by referring to Richard Clark who asserted that *technologies are merely vehicles that deliver content and teaching, but that by themselves they do not influence learning*. It is perhaps on this premise that Dagada emphasises in this book the importance of instructional strategies and assessment which should be built into computer-integrated learning.

The picture that the author paints and some of the key aspects he raises are those of interventions brought about by the South African government in its endeavours to address skills shortages in South Africa. Further, he goes on to assess the impact that these programmes have had and also the challenges they present. Dagada investigated the aggressive drive and introduction of Accelerated and Shared Growth

Initiative for South Africa (AsgiSA) and the Joint Initiative for Priority Skills Acquisition (Jipsa) and their impact on transforming human resources development through computer-integrated learning. In a nutshell, this book is about *Time, Space and Pace* with regard to learning. I have no doubt in my mind that readers will find this book to be both inspiring and educating.

David Redshaw
Chief Executive Officer
Bytes Technology Group
Johannesburg

FOREWORD

By Mthuli Ncube

Upon receiving the request to compose an foreword for this book, I decided to take the approach of ICT in education within the context of a developmental state. The discipline of development studies has this famous comparison of the economic history of Malaysia and Ghana. These two countries had similar post-colonial genesis, having attained political independence in 1957. Similarities end there because macroeconomic policies adopted by each country have led to exceedingly different outcomes. The dominant strand of analysis has always centred on politico-economic realms, celebrating Kuala Lumpur fiscal and central planning prudence and critical of Accra's dearth of vision, prioritisation of continental decolonisation project ahead of economic wizardry, and rampant military despotism.

My post-modernist interpretation of the Ghana-Malaysia comparison reveals that differences in the development of Malaysia and Ghana are attributable, in major part, to a fundamental factor that is currently inalienable in accelerating economic growth.

To start with, when both countries attained independence they were exceedingly characterised by acute poverty. Ghana's chances of a good launch were dashed when the charismatic founder Nkwame Nkurumah was deposed, heralding chaotic governance under military administrations until constitutionalism was restored in 1992. The current economic picture is a testimony of that false start: its GDP is at $60bn, per capita income at $2 700, an export-led dependent economy dominated by the agricultural sector, and aided by gold, timber and cocoa for earning foreign currency.

On the other hand Malaysia pursued the innovative economic policy Asia has given the world; the developmental state, the structure-ideology nexus that provides for an intervention of the state in the economy and pursuance of specialised industrial policy. In the case of Singapore, South Korea and Malaysia, the developmental state was observably marked by a concerted appreciation of the benefit of technology and ICT in education in the economy, thus explaining their current domination of knowledge-based industries.

It is this knowledge-based development trajectory that is attributable to the Malaysian economic miracle: its GDP currently stands at $308bn, per capita at $12 700, inflation at lowly 3.8% and GDP complemented by industry (48.1%), services (43.6%), and agriculture (8.3%) – signs of a thriving modern economy that has for years posted enviable growth rates.

In view of this, I therefore posit that the difference in the Ghana-Malaysia development trajectories, notwithstanding common genesis, is attributable in greater part to infusion of technology and ICT in education in the growth model. Efficient technologies increases productivity, lessen production costs, promote quality, increases competitiveness and innovation, leading to client satisfaction and growth. The same is true in the education front; by injecting technologies in the training environment, we accelerate efficiency in the delivery chain, increase quality in training, and promote technology transfer so inalienable in the contemporary world of global capitalism

Rabelani Dagada's *Time, Space and Place: Computer-integrated Learning in South Africa* – concerned with how ICT, and computers in particular, can aid the learning and training process – is exactly what the developing world needs right now. The majority of poor countries are characterised by a severe lack of skilled and expertise-laden human resource base, and are perpetually languishing in the backwaters of global prosperity. South Africa has not escaped this quagmire; it lacks skilled personnel in strategic industries such as artisans, engineers, specialised health professionals, numeric and natural science educators. As it has become quite clear, importing skills is not a reliable option. For South Africa and the affected world, the way out is a fast-tracked ICT-backed learning and training dispensation to offset the glaring gap.

Professor Mthuli Ncube
Head and Director of Wits Business School
University of the Witwatersrand

ACKNOWLEDGEMENTS

I would like to express my heartfelt gratitude and sincere thanks to my mentors, Professor Mariki Eloff (University of South Africa), Dr Maria Jakovljevic (Wits University), Professor Lucas Venter (University of South Africa), Dr Geoffrey Lautenbach (University of Johannesburg), and Professor Duan van der Westhuizen (University of Johannesburg). The guidance received from the aforementioned mentors enabled me to produce research that has exceeded any academic work I have done before. Whilst Dr Jakovljevic produced a better researcher out of me, Dr Lautenbach improved my writing tremendously. Professor van der Westhuizen is one of the pioneers of ICT-integrated learning in South Africa and thus he is my role model. In fact he taught and groomed me for many years.

Another distinguished academic who has decided to put me under his wings is Professor Mthuli Ncube, the Director of the Wits Business School (WBS). I was excited when he and Professor Louise Whittaker offered me a lectureship position. Professor Ncube came in handy when he composed the *foreword* of this book. He also encouraged me to introduce and direct a management development program in ICT for managers and executives. Wits University can be a very lonely and frustrating place for a novice academic. It is in this premise that I am thankful to the support that I received from the following academics – Tom Addison (School of Economic and Business Sciences), Professor Jason Cohen (School of Economic and Business Sciences) and Mark Peters (WBS). My academic area leader, Dr Terri Carmichael's unwavering support is awe-inspiring. She encouraged several MBA students to allow me to supervise their research work during my first year at the WBS. Terri, thanks for the vote of confidence. I felt honoured when Professors Mary Metcalfe and Ian Moll of Wits School of Education agreed that I should be part of the team that would help to introduce postgraduate courses in ICT in Education. Subsequent to this I was invited by Professor Ian Moll to participate in the PanAfrican Research Agenda on the Pedagogical Integration of ICT. Thanks for having confidence in me, Ian.

At the University of South Africa (Unisa) I was highly motivated and guided by Professor Wendy Kilfoil and that bunch of Learning Developers at the Institute for Curriculum and Learning Development (ICLD). In the ICLD, it was particularly Dr Maurice Vambe who encouraged me to write books. My broad understanding of ICT management issues were directly and indirectly honed by Dr Vick Coetzee, the Executive Director for ICT at Unisa. When I left Unisa, he gave me his doctoral theses and ICT strategy that he authored – I wish he knows that I use these publications as my bible for ICT management and leadership. Dr Wanjira Kinuthia, an Assistant Professor of Instructional Technology at the Georgia Sate University in Atlanta, USA, has been a pillar of strength for the past few years. We have researched the use of ICT for teaching and learning purposes in the South African Higher Education. Findings of this research were presented at an international conference and published in a journal. I wish I could acknowledge my counterparts in corporate environment, but this may constitute a clash of interest since most of them are my current or potential service providers.

I wish to thank the following practitioners, managers, executives and their organisations who participated in the study on which this book is based. These include the following: Tertia Albertyn (Safmarine Computer Services), Michelle Chandler and Douglas Mckay (Discovery Institute), Ravika Singh and Natalia Santos (Liberty Group), Frank Groenewald (Bank Seta), Zola Mkumla and Pieter Olwage (First National Bank), Chris Elfick and Hendra Bezuidenhout (Anglo American), Fiona Moncur (Deloitte & Touche), Wanda Minnaar and Marlon Maistry (Eskom), Charles Rosario Nunes and Gideon Vermaak (Telkom), Dr. Herman J van der Merwe (Tshwane University of Technology), Dinay Jansen (Vodacom), Paul Johnson (Learning Advantage), Sindy Zidel (Accenture), Martin Kopsch (Standard Bank), Prof Johannes Cronjé (University of Pretoria), Michelle Chandler and Douglas McKay (Discovery Institute), Dean Strooh (MTN), Andy Brown (Centre for Learning, Teaching & Development – Wits University), Amanda Jordan (Kumba Resources), Dr Isabeau Korpel, Lindi Lucas, Juliane Vorster, and Bev Judd (ABSA), Alison Jacobson (Reusable Objects), Johan Moller (Technikon SA), Sibongile Sambo and Willie Maritz (De Beers), Dr Jena Raubenheimer (eDegree), Dr Daleen van Niekerk and Dr Japie Heydenrych (Unisa), Keith Miller (Nedcor), and Aletta Michalopoulos (Westbank). At the time in which this book was heading for the printers some professionals were no longer working for the organisations that are reflected against their names. I should hasten to indicate that the names of people who were interviewed as part of group interviews do not appear in the above list. Nonetheless, their contribution is highly appreciated. The fieldwork was the most fulfilling part in the process.

On 1st February 2006, the time I resumed my duties as Manager for ICT and Knowledge Management at the Royal Bafokeng Administration (RBA), I was still in the process of writing this book. RBA is the infrastructural and service delivery arm of the Royal Bafokeng Nation (RBN). Due to its mandate, most commentators equate it to a municipality. The experience and exposure that I gained at the RBA has a direct positive impact in the fruition and content of this book. I am very thankful to the management of RBA for providing me with such an important opportunity of rolling-out ICT in the administration and the whole community. There are two women in RBA who made me smart – Khumo Molobye and Maphefo Magano. They are part of my team in the ICT and Knowledge Management and they helped me with the intellectual and "soft part" of my responsibilities. Any remarkable ICT Manager will tell you that the technical side would not succeed if the "soft issues" are not properly handled. When I leave RBA, I will walk into some big job; thanks to Maphefo and Khumo.

Although I am overwhelmed by the intellect and intimidated by the thoroughness of *Kgosi* Leruo Tshekedi Molotlegi, *His Majesty the King* of RBN, I was humbled by his personal support for our initiatives. *Kgosi* allowed my budget to be increased by more than 1 000%. The IT Managers in municipalities who had budgets bigger than mine were mainly those who were based in the metros – City of Johannesburg, City of Cape Town Metro, Nelson Mandela Metro, eThekwini Metro, Ekurhuleni Metro, and Polokwane. I was also highly motivated by Kgosi's passion for education and research. Niall Carroll, the Chief Executive Officer of the Royal Bafokeng Holdings (RBH) was very supportive towards the completion of this book. He would regularly ask me about the progress. Carroll and the team at RBH are also very helpful towards our ICT projects at the RBA and RBN. They would, for example, channel the Corporate Social Investment funding to some of the ICT projects. Thanks Niall. I am sincerely gratified by the leadership of the RBN, particularly *Her Excellency, the Queen Mother*, Dr SB Molotlegi and *Prince* Bothata Molotlegi for the solid support that they offered in my endeavour to roll-out ICT projects effectively. Other colleagues in the RBN who indirectly added value in my professional development are Ian Maclachlan, Susie Crossman and the whole team in the Royal Bafokeng Institute.

I am highly pleased with the professional and efficient manner in which staff at the Unisa Press contributed towards this book. They did a lot to make this book accurate. Any lingering errors and omissions in this book are solely my fault. The following Unisa Press's staff members added a lot of value in the shaping of this book: Sharon Boshoff, Charl D Schutte, Beth Le Roux, Liz Stewart, Lindsey Morton, Hetta Pieterse, and Elsabé Viljoen. I want to particularly thank Dirk van Enter who died whilst he was working on this book; may his soul rest in peace.

Eric Mudzanani, Mkhize, Avhapfani Tshifularo, Ezzy Lukhaimane, and Seth Mukwevho have been more than friends to me. I am humbled by the support they offered towards my academic career. There are times wherein they are fascinated by my obsession with books whilst other young black men are pursuing Black Economic Empowerment initiatives. Nevertheless, they allowed me to be who I am - a *bookaholic*. Most of my friends were very glad when I took a position out of the academia. Your friendship is highly appreciated; thanks.

I usually leave home at 7h40 and return at 23h00. This should be attributed to the long hours that I spend working. It is on this ground that I rarely visit my mother and extended family in Limpopo. I wish I could spend more time with my children who live with their mother in Gauteng, Naledzi and Tshidivhano. Now that projects have settled down I am able to spend every other weekend with my children. Although I love my family, work has made me a distant family member and an absent father. But still, I am grateful to my family – particularly my mother who put up with and supported me in my commitment to become the best. I will never forget where I come from. Yes, my father passed away when I was nine years old, but mama gave me the greatest gift – a decent upbringing and support to do whatever I wanted to pursue. My departed father would have enjoyed sharing the joy of this book with his youngest son. May you rest peacefully, dad.

To my partner Nkaba and our daughter Denga who have been more than patient during the course of composing this book: it has been a stretched and tiresome journey but you have made the travelling seem a lot easier at times. I am now looking forward to us exploring new routes together as a family. I have always been astounded by your beauty, grace and patience. I love you.

Rabelani Dagada, Jr
Phokeng, Rustenburg

To enable the reader to contextualise the content, arguments, findings and conclusions reflected in this book, it should be noted that it was preceded by extensive research that I conducted in the South African corporate training environment – under the auspices of the Faculty of Commerce, Law and Management at the University of the Witwatersrand in Johannesburg, South Africa. In my literature survey, I established that computer-integrated education, training and development is used both by institutions of higher learning and by corporate institutions to deliver and facilitate learning. However, the question arose whether computer-integrated learning was an acceptable alternative to face-to-face training in the corporate world.

Moreover, it remained to be established whether computer-integrated learning was providing any kind of solution to the human resources development challenges arising from South African legislation, including the South African National Qualifications Framework (1995), the Skills Development Act (1998), the Employment Equity Act (1998) and the Skills Development Levies Act (1999). I noted that very little research had focused on the use of computer-integrated learning in the South African corporate environment. The scarcity of literature in this regard seemed to indicate the need for research in this field.

Computer-integrated learning strategies and tools should be seen to add value in the corporate training environment in the South African context, but the literature did not show which computer-integrated learning tools and techniques were appropriate for human resources development.

The aim of the study on which this book is based was to assess the way in which computer-integrated learning was being used in the corporate training environment in South Africa. The study therefore reflects training and learning strategies, tools and techniques used and perceived as appropriate for the corporate training environment. I hoped to discover important insights into the future application of computer-integrated learning by South African companies, human resources development managers and policy formulators.

A qualitative approach was employed, with observation, legislation analysis and focus-group and key-informant interviews being conducted. The rationale for using the qualitative approach was that respondents constituted a rich and valuable source of information. My study took the form of a generic investigation of how 15 South African companies were using online learning to improve human resources development.

The participants in the research were corporate trainers, managers and trainees in 15 South African companies that used online learning as a training delivery mode. This study tried to involve companies in various industrial sectors including the energy, mining, insurance, banking, telecommunication and industrial services sectors.

By using multiple data-gathering methods and sources I aimed to satisfy the requirements of triangulation so as to ensure the validity of my study. Internal validity was applied by matching the research findings with reality in the corporate training environment. External validity was achieved by providing a sufficiently detailed description of the context of the study for the reader to compare the findings with other comparable situations.

This book also builds and extends on my extensive experience in the use of computers for training purposes. I have been involved in national projects wherein computer centres were established for teaching and learning purposes. During my tenure as the Manager of ICT and Knowledge Management at the Royal Bafokeng, I played a leading role in the establishment of e-Governance initiatives, in this instance constituted by Wireless Broadband Infrastructure, Voice Over Internet Protocol, Royal Bafokeng e-schools, Multipurpose Community Centres, and a Virtual Museum. Through this initiative *Dikgosana* (hereditary headmen), members of the executive council, and ordinary community members received computer training and access to technology. The knowledge and skills that I acquired in these multimillion-rand projects are infused into this book.

Rabelani Dagada, Jr
Phokeng, Rustenburg

INTRODUCTION

Several authors claim that computer-integrated learning will ultimately become the new training paradigm, taking its place alongside traditional contact situation training and changing the face of training generally. A number of trends – which include technological developments, the growth of the Internet, legislation and business imperatives – have accelerated the move to computer-integrated learning in the corporate world.

This book synthesises my experience, literature and research. Certain sections reflect analysed data derived from interviews in the South African corporate environment. I have also used case studies from various sources to contextualise certain assertions, claims, reports, arguments and findings.

Chapters 1 and 2 provide a brief overview of computer-integrated learning. These chapters also reflect the context of South African learning, training and development. The role of the Constitution of the Republic of South Africa and other legislation and imperatives in driving computer-integrated learning are dealt with. In chapter 2, where I claim that computer-integrated learning can play a major role in solving problems such as unemployment, the shortage of a skilled labour force, underdevelopment, poverty and other social ills.

Chapter 3 focuses on organisational issues that should be considered when implementing computer-integrated learning. I argue that computer-integrated learning should not be done in isolation from other business functions. My assertions are backed up by literature and analysed data from the South African corporate environment.

Chapter 4 addresses tools and strategies for computer-integrated learning, and shows how South African organisations employed various computer-integrated strategies and tools to transform education, training and development, as well as how certain factors in the South African corporate training environment affect learners' proficiencies.

Chapter 5 deals with issues related to the quality assurance of computer-integrated learning as required by South African legislation. This chapter reflects measures used by South African organisations to ensure the quality of computer-integrated courses.

Chapters 6, and 7 deal with computer-integrated assessment concepts, philosophical underpinnings and types of questions directed at learners.

Chapter 8 provides the final word and conclusion to the book.

While the study on which this book is based focused on the South African corporate environment, its findings may be transferable to other countries, institutions of higher learning and the public sector. My study appears to indicate that the South African organisations investigated were fairly competent in integrating computer-integrated education for human resources development in the corporate training environment. This could be attributed to the fact that managers and facilitators were using various strategies, techniques and computer-integrated tools in their endeavours to integrate online learning into their training environment. The aforementioned competence holds great promise for the future of levels of productivity in South African organisations. Thoughts and ideas expressed and recommended in this book will make a difference only if they are tested and implemented.

CHAPTER 1

COMPUTER-INTEGRATED LEARNING
AT A GLANCE

CONTENTS

1. INTRODUCTION AND BACKGROUND

Online learning (also known as electronic learning or computer-integrated education (CIE)) will eventually become the new training model, taking its place alongside traditional contact situation training and changing the face of training generally (Van der Westhuizen 1999:1). A number of trends, which include technological developments and the growth of the Internet, have accelerated the move to CIE in educational institutions and the corporate world (McLester 2002a:30; Westerman 2001:80). According to Trotter (2002:11), CIE will overtake conventional classroom training models over the next few years, claiming half of the overall corporate training market share. Technology and the Internet facilitate a more active role in the training process (Beffa-Negrini et al. 2002:334). The corporate world has realised that CIE is widely recognised as a viable vehicle for education in institutions of higher learning, as evidenced by the number of renowned academic institutions that now offer e-learning courses, including Harvard, Stanford and Pretoria universities (Van der Westhuizen 1999:1).

To date, little research has been conducted on the subject of the integration of CIE in the corporate training environment for human resources development. The contribution of CIE to human resources development as required by South African legislation should be investigated (Skills Development Act 1998). There is a gap in the literature as to which online tools and instructional and learning strategies are suitable. For the purpose of this book, human resources development refers to training, coaching and guiding in the workplace.

To ascertain the scope of the study on which this book is based, a number of searches were conducted to identify South African studies that relate to how CIE can be used by companies in South Africa to improve human resources development.

1.2 COMPUTER-INTEGRATED LEARNING

1.2.1 Key concepts and terminology

The terms 'online learning' and 'e-learning' will be used synonymously and interchangeably. According to Morrison (2003), online learning and e-learning mean the same thing. These terms, as used in this study, refer to any form of learning in which learning facilitators, learners and resources are sometimes separated from one another and where communication networks and computers provide the interaction between those people and resources (Alden 1998 in Lautenbach 2000:3). This separation should not be seen as a permanent feature in the e-learning environment because at times e-learning and traditional face-to-face learning are blended (Business Learning Institute 2003:10). The term 'computer-integrated education, training and development' will sometimes be shortened to 'computer-integrated education'. Computer-integrated education (or learning) refers to the use of computers for education, training and development. The computers in this instance could be connected (to each other or the Internet) or not. Online learning or e-learning is an element of CIE.

The terms 'generic content', 'off-the-shelf content' and 'third-party content' will be used synonymously and interchangeably. According to Morrison (2003), these terms refer to the same thing: courses or learning materials that have been designed and developed by content vendors for use in more than one organisation.

The terms 'face-to-face learning', 'traditional learning' and 'classroom learning' will also be used synonymously and interchangeably. According to Morrison (2003), they have the same meaning.

A few of these terms will now be listed to enable readers to understand e-learning technical terms. The list will enable the author and the readers to communicate in the same language because everyone has something to say about CIE.

- *Application*: Computer software; also referred to as a program. Many types of software fall under the category of application. Application software, however, is distinct from other software, such as operating system software and utility software.
- *Accreditation*: The process of approving the standards of education and training providers against criteria agreed by the South African Qualifications Authority (SAQA).
- *Assessment*: The process employed to systematically evaluate a learner's skills, knowledge, attitudes and values. It includes producing and gathering evidence of learner competence; evaluating the evidence against the learning outcome; and recording the results.
- *Assessment criteria*: The evidence needed to ascertain whether an outcome has been achieved.
- *Asynchronous learning*: Learning in which interaction between learning facilitators and learners occurs, occasionally with a time delay.
- *Bandwidth*: The information-carrying capacity of a communication network.
- *Blended learning*: Learning activities that combine aspects of online learning and traditional learning.
- *Broadband*: The ability of the communication channel to transmit data, audio, and video all at once over long distances.
- *Business requirements*: The condition an online learning solution should meet to align with the needs of stakeholders.
- *Computer-based education/training/learning*: A general term for the use of computers in both instruction and management of the training and learning process.
- *CD-ROM (compact disc read-only memory or compact disc-only media)*: A computer storage medium that can hold more than 600 megabytes of read-only digital information.
- *Chat*: Real-time text-based communication in the virtual learning environment. Chat can be used for learners' queries, instructor feedback or group discussion.
- *Content management system*: A centralised software application that facilitates and streamlines the process of designing, developing, testing, approving and posting online learning content on web pages.
- *Content*: The intellectual property and knowledge to be conveyed to learners. The formats of e-learning content include text, audio, video, animation and simulation.
- *Courseware*: Any kind of training or educational course delivered over the Internet.
- *Customer relationship management (CRM)*: Methodologies, software and Internet capabilities that help the enterprise to distinguish and group customers and to manage relationships with them.
- *Customisation*: The tailoring of course content to meet the specific needs of a geographical area, target audience or company.
- *Digital*: An electronic signal that can be transmitted faster and more accurately than an analogue signal.
- *Diagnostic assessment*: Enables the learner to recognise any knowledge or skills gaps when illustrating his/her present proficiency. Diagnostic assessment supports learning and advancement of the learner.

- *Discussion boards*: Forums on the virtual classrooms where learners and facilitators can post messages for the online community to read.
- *Distance education*: Education/learning situation in which the learning facilitators and learners are separated by time and location.
- *Education and training quality assurers (ETQAs)*: ETQAs are bodies whose responsibility is to ensure that learners receive qualifications only when they have satisfied the requirements of nationally agreed standards recognised by SAQA. They also accredit providers of education and training based on the criteria recognised by SAQA.
- *E-learning (electronic learning)*: A general term that covers a wide set of applications and processes, such as web-based learning, computer-based learning, virtual classrooms and digital collaboration. It includes the delivery of learning material via the Internet.
- *Electronic performance support system*: A computer application that is directly linked to another application to train or guide employees. This helps them to obtain information that enables them to accomplish a task.
- *Evaluation*: A systematic method of collecting information about the effectiveness and effect of a learning offering. Results of evaluation can be used to improve the course material or learning program. Evaluation includes determining whether learning outcomes have been achieved.
- *Face to face*: A term that describes the traditional classroom environment.
- *Facilitator*: The computer-integrated learning instructor who aids learning in the e-learning environment.
- *Feedback*: Communication between the facilitator or system and the learner resulting from an action.
- *Formative assessment*: A form of assessment that occurs continuously and enables learners to improve their performance while accumulating new competence.
- *Holistic assessment*: Holistic assessment focuses on the integration of outcomes and various assessment methods in order to gauge the learner's blend of knowledge, skills, values and attitudes.
- *Human resources development*: A term that describes the organised learning experiences such as training, learning, education and development offered by employers within a specific time frame to improve employee performance or personal growth.
- *Hypertext markup language (HTML)*: A programming language used to create documents for display on the Internet.
- *Hypermedia*: Applications or documents that contain dynamic links to media, such as audio, video and graphics.
- *Instructional designer*: An expert who applies a systematic methodology based on instructional theory to create content for learning.
- *Interactive media*: It allows for two-way interaction or exchange of information among the online learning participants.
- *Intranet*: A local area network (LAN) or wide area network (WAN) that is owned by a company, accessible only to employees, and secured from outside intrusion by a combination of firewalls and other protection measures.
- *Knowledge management*: The process of capturing, organising and storing information and experiences of employees and groups within an organisation and making them available for employees. This helps a company to gain competitive advantage.

- *Learning content management system*: A software application that manages the development, storage, use, and reuse of learning content.
- *Learning solution:* Any combination of technology and method that delivers learning, including software or/and hardware products that vendors tout as solutions to enterprise learning needs.
- *Learning management system:* Software that automates the administration of learning. It registers users, tracks courses, records data from learners and facilitators, and provides management with reports. A learning management system is designed to handle courses by several publishers and providers. In most instances it does not have its own authoring capabilities, but it deals mainly with the management of courses created by a variety of other sources.
- *Mentoring*: A career development process in which less experienced employees are matched with more experienced employees for guidance.
- *Multimedia*: A term that covers interactive text, images, sound and colour.
- *Online community*: A meeting place for people on the World Wide Web. It is designed to facilitate communication and cooperation among people who share common interests and needs.
- *Online learning*: Learning that is delivered by web-based or Internet-based technologies.
- *Outcomes*: The knowledge, skills, values and attitudes that a learner should be able to demonstrate after a particular learning experience.
- *Portfolio*: A collection of learner's projects completed over a period of time. Portfolios should demonstrate the learner's command of particular knowledge and skills and be able to reflect the learner's attitudes.
- *Self-assessment*: The method by which the learner determines his/her own level of skills and knowledge.
- *Self-paced learning*: A learning method according to which a learner determines the pace and timing of content delivery.
- *Skills gap analysis*: An analysis that compares an employee's skills with the skills required for the job he/she has been given.
- *Subject matter expert*: A person who is recognised as having sufficient knowledge and skills in a particular subject area.
- *Soft skills*: Business skills such as communication and presentation, leadership and management, human resources, sales and marketing, professional development, project and time management, customer care, team building, administration and personal development.
- *Synchronous learning*: A real-time learning, facilitator-led online learning event in which participants are logged on at the same time and communicate directly with one another.
- *Summative assessment*: A form of evaluation that assesses the competence of the learner over a period of time, for example after completing a chapter or at the end of the semester.
- *Teleconferencing*: Two-way electronic contact between two or more groups in separate locations via auditory, video and/or computer systems.
- *Virtual*: The term means not concrete or physical, because there are no buildings, and interaction is done over the Internet.
- *Virtual classroom*: The online learning space where learners and facilitators interact.

1.2.2 The benefits of computer-integrated learning

Many authors have discussed the ways in which online learning can be used for the delivery of training, assessment and support (Kwinn 2001:23; Fichter 2002:70). CIE offers a variety of possibilities in terms of training, ranging from highly complicated flight simulation to basic drills and practice, from video conferencing to tutor support across an electronic mail (e-mail) link, and learning over the information super-highway using a stand-alone personal computer (Beffa-Negrini et al. 2002:335).

CIE has made considerable progress since the early 1980s (Lautenbach 2000:16), attributable in large measure to technological developments. Technological improvements have been fast, and so have the changes in corporate training methods. CIE as a corporate training method has been enhanced by virtuality, which now manifests itself in aspects such as content provision, electronic access to libraries, e-books, discussion rooms and chat lines (Abell and Foletta 2002:396). Soon these channels will become mainstream delivery modes for corporate training and part of the competitive advantage of a successful company (Christner 2003:130; Jefferson 2000:54).

A significant benefit of computer-integrated learning is that it allows learners access to learning material at their convenience (De Lima 1999:29). The advantage for the corporate world lies in the fact that training can be offered without the necessity for a physical classroom (Beffa-Negrini et al. 2002:334), as learners can learn wherever there is access to the Internet (Taylor 1999:10). Moreover, the interactivity inherent in Internet-based courses has fuelled the growth of online learning (Chambers 2002:9). Trotter (2002:12) notes that learners who use the Internet have reported greater engagement in the learning experience than in the more static learning associated with the traditional classroom. Taylor (2002:11) writes that CIE enables the instructor to monitor the learner's progress continuously. Learners become involved in the learning process, and modules can be designed to suit different learning styles (Hamilton-Pennell 2002:12).

Arnone (2002a:34) and O'Connell (2002:15) report that some learners find that online learning suits their learning styles better than the conventional, face-to-face options because they are more visual than auditory. Furthermore, some learners prefer to work at their own pace and not to restrict their learning to a specific location (Hamilton-Pennell 2002:3).

Christner (2003:130) remarks that CIE offers learners a range of options for navigating through the lessons, submitting assignments and holding discussions with other learners. In most instances, discussion takes place through a threaded discussion feature, in which learners can send messages on a specific topic (Beffa-Negrini et al. 2002:336). Learners may also respond to messages posted by the trainer. Content may include graphics, tables, screen shots, illustrations and multimedia elements.

1.2.3 The shortcomings of computer-integrated learning

Hamilton-Pennell (2002:3) cautions that both critics and supporters have identified some weaknesses associated with online learning. These include the lack of social presence usually associated with physical classrooms, because learners miss the real-life interaction with their colleagues and the instructor (Bridges 2000:50). Taylor (2002:11) considers that this feeling of loneliness could be a serious stumbling block to learning; in adult education in particular there is much that learners can learn from each other. Christner (2003:130) observes that it takes a long time for trust to develop among online learners.

The dilemma is that the very technology that makes CIE possible can constitute a hurdle, because online learners need certain types of hardware, technical support and fast Internet connection (Abell and Foletta 2002:397; Hamilton-Pennell 2002:3). Live discussion can be difficult if more learners are involved. According to Chambers (2002:8), learners may also struggle with the discussion thread software.

Some authors are suspicious of the quality of online learning materials (Mioduser, Nachmias, Lahav and Oren 2000:61). Subjecting CIE materials to quality standards, accreditation and legislation could solve this (Wessels 2001:221). In South Africa, learning materials must comply with the principles and requirements of outcomes-based education and training (OBET), the National Qualifications Framework (NQF) and the South African Qualifications Authority (South African National Qualifications Framework 1995). On the other hand, CIE should enable corporate employers and trainers to meet the requirements of labour legislation (Skills Development Act 1998; Employment Equity Act 1998; Skills Development Levies Act 1999). Employers should also interact with sector education and training authorities (Setas) with regard to training and human resources development.

Trotter (2002:12) and D'Amico (1999:64), however, feel that these challenges are not insurmountable, and that one way of dealing with them is through the provision of good instructional strategies and technical support to online instructors. Corporate instructors should be exposed to a program that will equip them with online instructional design. The Internet, too, has a number of tools and resources that can help instructors and learners.

1.3 THE CONTEXT OF EDUCATION, TRAINING AND DEVELOPMENT IN SOUTH AFRICA

1.3.1 A new democratic dispensation

Before the new democratic dispensation in South Africa in 1994, education and training were separated. Education was associated with academic qualifications, while training was associated with vocational qualifications. Academic qualifications were obtained in formal learning institutions, whereas vocational training was relegated to the so-called technical colleges and the workplace. At present, learners can access both academic and vocational qualifications via different routes. The qualifications obtained in formal learning institutions and the workplace are supposed to meet the national unit standards and have equal status in the NQF (IEB 2005:60). This setup should be attributed to the Constitution of the Republic of South Africa (1996) and the legislation (above). These acts were intended to foster close working relations between the Department of Labour (DoL) and the Department of Education (DoE). There has been some friction between these departments, however, owing to their fighting for turf.

The tension has been there since 2001. This was revealed by the minister of labour, Membathisi Mdladlana, who attacked the DoE for sluggishness in its part of the National Skills Development Strategy: 'I am frustrated as a minister of labour. We have to link education with training – what is frustrating is when you can't help because you train people and they don't know what button to push; people are in need' (*Mail and Guardian* 2004). The minister's frustration was owing to the difficulty in reviewing the NQF with the DoE. (The NQF is a mechanism for setting standards for all learning programs to ensure flawless link between training and education.)

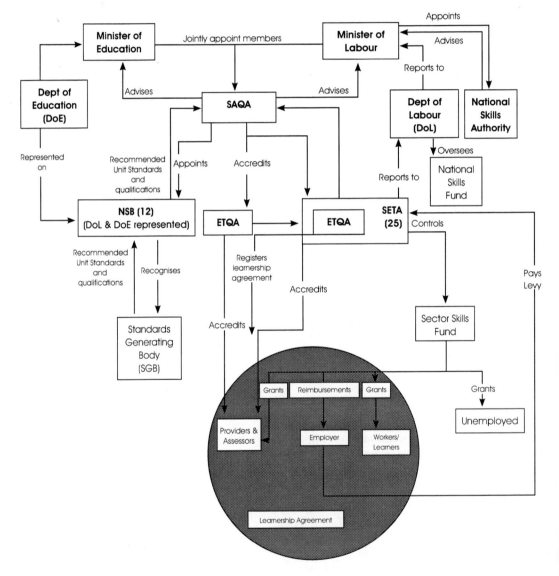

FIGURE 1.1 SOUTH AFRICAN EDUCATION, TRAINING AND DEVELOPMENT
Source: IEB (2005)

The difficulty of reviewing and implementing NQF should be attributed to the fact that the process involved two national departments. This was confirmed by the minister of labour who confessed that things would have been easier if education and training had been 'under one roof' (*Mail and Guardian* 2004). The failure to implement NQF not only affected the two departments but frustrated the whole education and training sector. In July 2001 the DoE assigned a task team to review the implementation of the NQF, which was completed in April 2002. This was followed by public comments and the establishment of the interdepartmental task team that produced a consultative document in 2003. However, since then there has not been interaction between the two departments, as reported in the *Mail and Guardian* (2004). In other words, the review of NQF implementation entered the fourth year in 2004 without any substantial progress. The improvement of education and training under the new democratic dispensation has its own difficulties. One should hasten to say, though, that there is much more good than bad. With more effort and determination, most of the stumbling-blocks will be resolved.

The main tenet of this book is that CIE is transforming education, training and development by enabling organisations to meet the requirements of the Constitution and the relevant acts. The next chapters will show how, for example, South African organisations use computers to educate, train and develop many people in different places simultaneously. Figure 1.1 provides a broad picture of South African education, training and development. It is interesting, as the figure shows, that the DoE and the DoL are working together to transform human resources through the establishment of various laws and structures.

The NQF (1995) brought about major changes in South African education, training and development. One is tempted to describe these as revolutionary. According to the IEB (2005:42), the NQF is a framework for legislation policy with regard to education, training and development and predetermines pathways and levels for progress in education and training. It also sets out principles for education and training. The NQF is administered by SAQA, which monitors and guides education and training providers.

Here are the principles of the NQF (IEB 2005:43):
- *Integration of education (knowledge) and training (skills)*: Education and training are seen as connected to each other rather than as separate entities. An outcomes-based system focuses on both education and training as it emphasises the interdependence between knowledge and skills.
- *Progression (achieving credits)*: The transfer of credits from different institutions, or between the world of work, education and training, is possible with an integrated qualification structure.
- *Portability (transfer of credits)*: Credits are accepted wherever they have been achieved.
- *Articulation (movement between education and the work environment)*: Synchronisation between education and training and the workplace is ensured by clear links between all service providers.
- *Flexibility*: Multiple entry and exit points are part of the NQF structure and will ensure that learners are able to enter and exit at different points.
- *Recognition of prior learning (RPL)*: Learners at different levels will be given an opportunity to show the knowledge, skills and values they possess and match these with a proposed field of study. They may be exempted from doing some tasks that

fall under the proposed field of study and concentrate only on those they have not mastered. A learner can also be recognised for his/her experience in relation to standards and qualifications if he/she is assessed as competent.

• *Democratic participation*: Education and training, especially assessment, has to be democratic in nature and should involve all stakeholders so that it is transparent.

• *Quality and credibility*: Qualifications obtained in South Africa have to be comparable with international standards, and this can be achieved only with a unified qualifications authority such as the NQF.

These principles are meant to merge education and training.

1.3.2   The operation of the NQF

The NQF sets out legislative parameters for structures that are involved in education and training. In fact, the NQF operates through these structures. These include the SAQA Act (1995), education and training quality assurers (ETQAs); and national standards bodies (NSBs). The operation of the NQF was also brought into effect by the Skills Development Act (1998) and Skills Development Levies Act (1999).

Figure 1.2 reflects how the NQF operates.

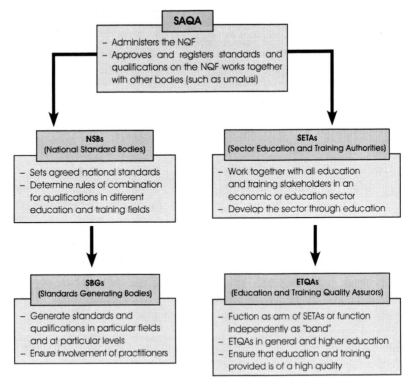

FIGURE 1.2   OPERATION OF THE NQF
Source: IEB (2005)

The DoE and the DoL adopted an outcomes-based education and training (OBET) system. OBET's purpose is to develop the knowledge, skills, attitudes and values of learners. On this premise, SAQA implemented a number of critical learning outcomes that should be an integral part of any learning materials. These critical outcomes are to:

- Identify and solve problems in which responses display that responsible decisions using critical and creative thinking have been made
- Work effectively with others as a member of a team, group, organisation or community
- Organise and manage oneself and one's activities responsibly and effectively
- Collect, analyse, organise and critically evaluate information
- Communicate effectively using visual, mathematical and/or language skills in the modes of oral and/or written presentation
- Use science and technology effectively and critically, showing responsibility towards the environment and health of others
- Demonstrate an understanding of the world as a set of related systems by recognising that problem-solving contexts do not exist in isolation

Education and training are seen as connected to each other rather than as separate entities. An outcomes-based system focuses on both education and training, as it emphasises the interdependence of knowledge and skills. To contribute to the full personal development of each learner, and the social and economic development of society, the intention underlying any program of learning must be to make an individual aware of the importance of the following:

- integrating education (knowledge) and training (skills)
- reflecting on and exploring a variety of strategies to learn more effectively
- participating as responsible citizens in the life of local, national and global communities
- being culturally and aesthetically sensitive across a range of social contexts
- exploring education and career opportunities
- developing entrepreneurial abilities

1.3.3   Human Resources Development Strategy for South Africa

In 2001 the South African government's Development Strategy Cluster (composed of the Ministries of Education, Labour, and Arts, Culture, Science and Technology) presented the Human Resources Development Strategy (HRD) for South Africa (Asmal 2001).

a) A nation at work for a better life for all

The HRD supported President Thabo Mbeki's vision of 'a nation at work for a better life for all'. The mission of this strategy is to maximise the potential of the people of South Africa through the acquisition of knowledge, skills and value, to work productively in order to achieve a rising quality of life for all, and to set in place an operational plan, together with the necessary institutional arrangements, to achieve this. The HRD and its mission impact significantly on the strategic direction and operation of education, training and development in the

South African context. One can argue that this strategy is one of the drivers of CIE. Education and training practitioners have employed CIE as one of the delivery modes that could enable them to transform South African education, training and development as required by the strategy.

b) Overarching goals of the Human Resources Development Strategy

According to the then minister of education, Professor Kader Asmal (2001), the strategy has three overarching goals:
- To achieve an improvement in the UNDP human development index for South Africa, as a result of improvements to the social infrastructure
- To reduce disparities between the rich and the poor, reflected in an improved 'Gini co-efficient' rating
- To improve the country's position in the International Competitiveness League

c) Strategic objectives of the Human Resource Development Strategy

To achieve these goals, there are five strategic objectives:
- To improve the foundations for human development
- To improve the supply of high-quality skills, especially scarce skills, which are more responsive to social and economic needs
- To increase employer participation in lifelong learning
- To support employment growth through industrial policies, innovation, research and development
- To ensure that the above four initiatives are linked

Figure 1.3 reflects how the HRD should create a linked system by integrating various elements to make sure that human resources correspond with the needs of society and the economy.

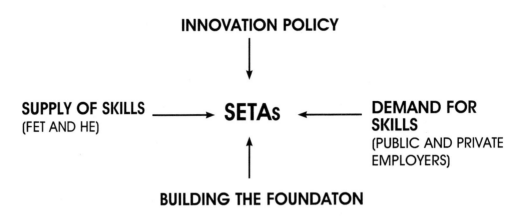

FIGURE 1.3 SOUTH AFRICAN HUMAN RESOURCES DEVELOPMENT STRATEGY
Source: IEB (2005)

1.3.4 Accelerated and Shared Growth Initiative for South Africa

a) Poverty and unemployment reduction

In 2004 the South African government made a commitment to reduce poverty and unemployment by 50 per cent by 2014 (Tang 2006). This determination was boosted by the impressive economic growth which had exceeded 5 per cent in 2005 (Wilson 2006). This kind of growth gave hopes for job creation. On the other hand, the rate of direct foreign investment was growing steadily. At the same time, the government has realised that the goal of decreasing unemployment to below 15 per cent and poverty by half will not be possible without strategic economic interventions by the government and other stakeholders such as labour and business (Tang 2006).

The deputy president, Phumzile Mlambo-Ngcuka, was given the responsibility of coordinating this strategic intervention (Mbeki 2006). That was preceded by the unveiling of the Accelerated and Shared Growth Initiative for South Africa (AsgiSA) by the cabinet in July 2005. The deputy president then put together an AsgiSA task force that included the 'ministers of finance; trade and industry; public enterprises; the premiers of Gauteng and Eastern Cape provinces, the mayor of Johannesburg, who represented the South African Local Government Association' (Asgisa 2006). The government wanted AsgiSA to be a national shared strategy rather than just a government programme.

b) Targets of accelerated and shared growth

Aspects targeted for the accelerated and shared growth include (Asgisa 2005):

- Infrastructure investment
    - Plans to be implemented in electronic communications
    - Provincial infrastructure

- Sector strategies
    - Further priority sectors
    - Cross-cutting industrial policy challenges that are also being addressed

- Educational and skills development
    - Educational response to the skills challenge

- Eliminating the second economy
    - Leveraging the first economy
    - Expanding women's access to economic opportunities
    - Measures to promote youth development during 2006/07
    - Leveraging components of broad-based black economic empowerment
    - Decisions on the small business regulatory environment
    - Measures to realise the value of dead assets

- Macro-economic issues
- Governance and institutional interventions

c) Education and skills development

AsgiSA is a critical element of skills, training and development. The government realised that the single greatest impediment to public infrastructure acceleration and private sector investment growth is the shortage of skills, 'including professional skills such as engineers and scientists; managers such as financial, personnel and project managers; and skilled technical employees such as artisans and IT technicians (Asgisa 2006; Mlambo-Ngcuka 2006). Among other things, the government perceives the skills shortage as a consequence of apartheid policies and the slowness of our education and skills development systems in catching up with the current aggressive acceleration of economic growth (Le Roux 2006).

AsgiSA (2006) intended to improve the skills pool by:
- Closing the gap between potential employers and employees. This will be done by establishing an employment services system and phase 2 of the National Skills Development Strategy
- Developing a scarce skills database based directly on the expected needs of over 100 individual projects included in AsgiSA
- Deploying experienced professionals and managers to local governments to improve project development, implementation and maintenance capabilities
- The Development Bank of Southern Africa (DBSA) will deploy an estimated total of 150 expert staff, with the first 30 to be deployed in April 2006. The project will include skills transfer to new graduates. The DBSA is compiling a database of 'retired experts' for this and further possible deployments.

d) Education and skills development: Joint Initiative for Priority Skills Acquisition

The Joint Initiative for Priority Skills Acquisition (Jipsa) is a new AsgiSA institution (Ntuli 2006). Jipsa is led by a committee comprising the deputy president, ministers, business leaders, trade unionists, experts and education and training providers. Jipsa has been given the task of identifying urgent skills needs (Mbeki 2006). The deputy president remarked that Jipsa is one of the building blocks of AsgiSA because, 'if we in the human resources and skills development sphere fail, AsgiSA will fail' (Mlambo-Ngcuka 2006b). Jipsa's mandate include formulating quick and effective solutions such as special training programmmes, bringing back professionals who have retired and Africans in diaspora. To fast-track the development of the trainees, Jipsa may send them overseas for mentoring and placement (AsgiSA 2005).

Jipsa will not replace the functions of existing learning institutions and Setas, but will focus only on scarce and critical skills without which South Africa will not deliver on AsgiSA targets (Blaine 2006; Jackson 2006). Nevertheless, Jipsa will relate to the schools and higher learning institutions, which were given broader mandates. Jipsa will achieve its mandate by, among other things, focusing on certain aspects (Mlambo-Ngcuka 2006):
- Increased efforts to support maths, science and English language skills in schools. Jipsa will focus specifically on teachers of these subjects.

- Jipsa will support the alignment of further education and training (FET) colleagues and higher education institutions in their work of producing graduates who meet the requirements of the public and private sectors.
- Jipsa will work with higher education institutions and employers, all of whom are represented in it.
- Jipsa will indirectly support the DoE's work in adult basic education and training (ABET). This will assist in drawing adults who are not skilled into the economy.
- Jipsa will work with the Department of Home Affairs (DoHA) to look at facilitating the importation of scarce and priority skills to enable AsgiSA to meet short- to medium-term skills demand.

Jipsa has several working areas that will enable it to achieve its goals. Based on AsgiSa priorities, certain working areas have been identified (Mlambo-Ngcuka 2006):
- High-level, world-class engineering and planning skills for the 'network industries', transport, communications and energy all at the core of our infrastructure programme
- City, urban and regional planning and engineering skills desperately needed by our municipalities
- Artisan and technical skills, with priority attention to those needs for infrastructure development
- Management and planning skills in education, health and municipalities
- Teacher training for mathematics, science, ICT and language competence in public education
- Specific skills needed by the AsgiSA, sectors starting with tourism and business process outsourcing (BPO) and cross cutting skills needed by all sectors especially finance; project managers and managers in general
- Skills relevant to local economic development needs of municipalities, especially developmental economists

Jipsa will assist in leapfrogging South Africans who are in the second economy into the first economy (Shezi 2006). The big boost in its work will be to empower people through education, training and development.

CHAPTER 2

RATIONALE FOR COMPUTER-INTEGRATED LEARNING

CONTENTS

2.1 INTRODUCTION

Organisations are faced with daunting challenges when taking decisions on the use of computers in training and learning. Can they use the Internet effectively for education and human resources development? They tend to wonder whether CIE can deliver training and learning of a quality that matches or even surpasses that of conventional face-to-face learning environment. Well-established organisations are evaluating their training missions and searching for other ways of providing training and learning (Collis and Moonen 2001:37). They want to deliver quality lifelong learning to many employees without being hindered by time, place and socio-economic status (McLester 2002a:24). To achieve this objective, organisations are shifting from the traditional emphasis on classroom training to an environment where learning could take place at any time and anywhere through computers (Capper 2001:238).

A history of online learning and the implications of the knowledge economy are detailed in this chapter. The main purpose is to reflect on the rationale for CIE in the corporate training environment. There are many reasons to support the use of CIE in South Africa. These include the shift in the workplace, meeting the requirements of the legislation, the introduction of sector education and training authorities, skilled workforce shortage, high level of brain drain, difficulties experienced by small businesses, unemployment, HIV/Aids pandemic, and an increasing poor white problem.

2.2 A BRIEF HISTORY OF ONLINE LEARNING

The emergence of CIE has been described as an 'accidental revolution' (Molnar 1997:63). This should be attributed to the innovation of computers, which have created the most stimulating ideas in the field of training and learning. Several factors and developments have driven the emergence of CIE:
- the global economy
- the scientific information explosion
- the emergence of cognitive science
- new educational demands

2.2.1 The global economy

Information and communication technology (ICT) led to the rapid movement of goods and resources and created globalisation through the interdependence of countries' economies (Molnar 1997:64). To gain maximum benefit from global markets, countries should be competitive and to accomplish this, they must have a well-trained workforce (Capper 2001:237).

Information-based organisations are emerging in which corporate human capital and knowledge are the most important drivers (Molnar 1997:66). The essence of this development is that companies need knowledgeable workers with a good understanding of ICT who can compete globally (Collis and Moonen 2001:192).

2.2.2 The scientific information explosion

The world is experiencing an unprecedented information explosion (Aggarwal and Bento 2000:2). Scientists and engineers use the Internet to access millions of databases that store statistics, numbers, data, maps, and chemical and physical structures (Molnar 1997:67). Collis and Moonen (2001:192) have noted that the volume of new information is growing rapidly. Changes in many sectors are making employees' basic knowledge and skills dysfunctional. Knowledge concepts and theories are constantly being revised. New theories are emerging, and the data is continually being modified. Fields of study are merging, and hyphenated sub-fields are being formed.

2.2.3 The emergence of cognitive science

There has been a shift in the education field from theories of 'learning' to 'cognition' (Molnar 1997:65). Cognitive science approaches training and learning differently. It deals with how a person processes and uses information rather than memorises facts. The learner should therefore be able to deal with learning constructively by engaging in critical thinking and applying problem-solving skills (Crane 2000:41). Many authors have noted the link between ICT and cognitive skills. The cognitive approach is significant in the training and learning environment because it recognises the person's information processing abilities and limitations (Molnar 1997:67). According to Crane (2000:41), ICT can assist a person by organising information to fit his/her capacity. ICT has therefore changed the emphasis in education from memorising theory to applying knowledge and skills.

2.2.4 New educational demands

Training and learning will eventually reflect the characteristics of the new economy that are shown in other fields of society (Collis and Moonen 2001:194). The impact of ICT took a decade to manifest itself in the commercial world and will also take some time to manifest itself in the training and learning sector (Capper 2001:237). All the same, the use of computers can be seen in thousands of learning institutions and organisations worldwide that are offering courses through CIE (Charp 2002:8).

Working people need learning that can take place anywhere and at any time, and thus the use of online learning is expanding worldwide. Charp (2002:8) estimates that online learning in the corporate sector will grow from US$2,2 billion to $18,5 billion by 2005. This may be because organisations are experiencing training budgets cuts, decreased interest in travelling to training venues, with concomitant costs of airplane flights and time away from home. Computer-integrated learning enables learners to learn conveniently. Charp (2002:9) outlined some examples of this convenience:
* Cisco Systems uses online learning to work with its sales force. At the recent Comdex meeting, Cisco's CEO, John Chambers, stated that 'e-learning is the next killer application'.
* McDonald's trainers log on to Hamburger University for additional training and updated information.

- Circuit City, with its 600 stores and approximately 50 000 employees, uses customised courses that they say are 'short, fun, flexible, interactive and instantly applicable on the job'.
- Microsoft is collaborating with the Oregon Graduate Institute of Science and Engineering at Oregon Health and Science University, so that workers can obtain a master's degree in software and technology management.
- CIE enables American soldiers to study while stationed in other countries.

The growth of CIE in the corporate environment could be attributed to:
- Increasing demand for equitable education for employees
- Providing training to employees whose needs cannot be met through the formal educational process
- Desire for organisations to be involved in online learning and to compete globally
- Additional profit owing to the performance of competent employees (Charp 2002:9).

CIE has become a mainstream training and learning method in most countries, in most learning institutions in developed countries, and in a large number of corporations (Capper 2001:237).

Later chapters will show how organisations in South Africa are using CIE for human resources development.

2.3 SOUTH AFRICA AND THE KNOWLEDGE ECONOMY

2.3.1 Dooming the underprivileged

ICT-enhanced learning should be introduced in South African organisations to tackle the technological imbalance. We are living in an age when most desirable jobs are technology-oriented. Learners with adequate computer exposure will enjoy better jobs, attracting fringe benefits and a superior lifestyle (Wilcox 1997:6). Employers therefore have a responsibility to ensure that all working people have access to computers, regardless of their ethnic or socio-economic background. Failing to expose working people (particularly black Africans) to computers is to doom the underprivileged to an underprivileged future (Den Biggelaar 1997:172). This will also shatter hopes for the success of the African renaissance.

2.3.2 The emergence of the knowledge economy: a new concept

The emergence of the knowledge economy is compelling companies to turn to knowledge management as a means of guiding their organisations and achieving growth (Wiig 1993:37). In South Africa the knowledge economy is still a new concept and perhaps a frightening one (Wiig 2000:3). Knowledge as such is not enough. If it is not mobilised and applied at the right time and place, it is useless (Van der Spek and Spijkervet 1997:35). What distinguishes the expert from the novice is the amount of knowledge he/she has accumulated and the ability to apply it effectively. South African organisations need employees with appropriate skills to amass incredible amounts of data (Den Biggelaar 1997:170). Workers who have undergone a process of learning should apply their knowledge and skills to mould data into information or knowledge that is able to bring benefits to the company. According to Cronjé and Baker (1999), a company should have technological support, because without it, corporate knowledge is lodged only in the heads of employees. When they leave the company, their knowledge leaves with them. With a technological support system, the corporate knowledge resides in the knowledge management applications and is available to any of its users (Wilcox 1997:10).

2.3.3   The third wave

According to Honey (1999), the first wave started 10 000 years ago when pastoral people turned to crop cultivation for survival. The age of agriculture brought stability and encouraged innovation as people stayed together in villages. The second wave began in the 18th century with the Industrial Revolution in Europe. Peasants and farm-workers began to leave their farms to work in industries, which gave birth to cities. The third wave is the current age of ICT and the knowledge economy (Grant 2000:29).

2.3.4   Fundamental change

The main difference between the second wave (industrial) and the third wave (infotech) is the mode of production. In the second wave, factories are static and raw materials must be brought to the loom, woven into fabric and then sold. The third wave, on the other hand, is dynamic; it is not confined by geography, bricks and mortar, and its primary asset is information and knowledge, which transcend boundaries (Honey 1999).

2.3.5   Towards the fourth wave

South Africa is still mainly in the first wave (the primary sector of agriculture, mining, fishing and forestry). In other words, the country still has a large, essentially peasant population in the rural areas and an economy based on raw materials (Cronjé and Fourie 2003:21). It is not a fully fledged industrial country (second wave). This state of affairs should be attributed to the lack of adequate labour skills. We should expose workers to ICT-enhanced learning so that they can accumulate skills that will move us to the third wave (Rolland and Chauvel 2000).

If South Africa fails to develop its people, regardless of their history, then the country is guaranteed to stay at the bottom of the heap. South African organisations therefore have the huge task of developing the second wave component of the economy to create jobs. We should also devote attention to building up our electronic infrastructure and plugging into the contemporary knowledge-based economy. If we do not, we will be trapping ourselves in perpetual backwardness (TFPL 1999). Basing our economy on the export of raw materials is a long-term losing strategy, even though there are short-term benefits.

2.3.6 Two waves of change

South Africa will benefit from the fast recovery of the Asian economies because their growth will be accompanied by high labour costs that will drive industrial investment in Africa. It should be noted, however, that South African labour laws are perceived as rigid and possibly repelling investors.

The government is perceived to be interfering in the running of the business sector at macro level. According to Swanepoel, Erasmus, Kirsten and Holtzhausen (2003:64), the protection of the commercial sector against government interference can be a source of strain on any economy. These authors went further and claimed that South African government intervention in the business sector leads to an increment in production costs and the closure of small enterprises. This will consequently lead to redundancies and retrenchments.

Swanepoel et al. (2003) use the McKeever Institute of Economic Policy to support their claim:

> Several costs imposed on South African business hurt its competitive edge. This is a partly because of legislation pushed through by the large trade unions, Congress of South African Trade Unions (COSATU), National Union of Mineworkers (NUM), National Council of Trade Unions (NCTA), United Workers Union of South Africa (UWUSA), combined these unions represent over two thirds of the country's (sic) labour force. This has given them large political clout enabling them to demand strong worker rights that have been passed in the form of Labour Relations Act, the Unemployment Insurance Act, Basic Conditions of Employment Act and the Insolvency Act. All of these pieces of legislation have placed huge additional costs on business.

If South Africa moves into the second wave, this investment must be channelled towards electronic third-wave development. South Africa will then have two waves of change rolling at the same time. The rural-agrarian economy should be improved to leapfrog the third wave. Telkom and Eskom should speed up their efforts to provide telecommunications and electricity services so that every village in the country may be online and every family have access to an Internet connection (Honey 1999).

2.3.7 Changing or becoming obsolete

South African organisations should help the government to equip communities and learning centres with computers and Internet connections. Failure to do so will have a negative impact, because organisations will not be able to change without skilled employees. They have to change or become obsolete. Nearly everything is becoming computerised. One of these days, people will be wearing computers (Taylor 1999). In homes, silicon chips control everything from dishwashers to microwave ovens and television sets. Over the past years, the transformation brought about by ICT in our homes, organisations and workplaces have been amazing. It seems 'we have not seen anything yet'. Many services that now require a physical touch will be served remotely over high-speed digital communications links (Taylor 1999). Organisations should prepare their workforces for this.

2.3.8   Computerised human body

Honey (1999) claims that the traditional keyboard is likely to be replaced in the long term, but significant advances in voice technologies will soon enable anyone to hold a conversation with a computer. In the next 20 years, computers may well become able to sense changes in their users' moods, for example by analysing facial expressions or voice tones.

2.3.9   Closing a huge knowledge gap

South Africa was isolated from the global economy for many years because of apartheid. Although the isolation was justified, it resulted in a huge knowledge disconnection for South African organisations from world-class business practices. As we re-enter the global economy, a huge knowledge gap should be closed through an effective knowledge capture, development and deployment process so that South African organisations become and remain internationally competitive (Honey 1999). The effective way of closing this gap is by expanding education, training and development. Equipping the labour force with skills will enhance the African renaissance and break the back of poverty in South Africa.

2.3.10 Irreplaceable capital

The only irreplaceable capital a country can possess is knowledge and the ability of its citizens to apply it (Taylor 1999). The productivity of that capital depends on how effectively workers share their knowledge with those who can use it. Whereas the South African industrial organisational framework is characterised by capital and labour, the new knowledge will be organised as a set of continuously shifting intellectual networks. Whereas the industrial focus is on physical assets such as production machines, the knowledge economy will be driven by intangible values.

2.4 RATIONALE FOR COMPUTER-INTEGRATED LEARNING IN THE CORPORATE TRAINING ENVIRONMENT IN SOUTH AFRICA

2.4.1   Shift in the workplace

According to Van der Westhuizen (1999), another challenge facing South Africa is the dramatic shift in the workplace that will result from the computerisation of functions that people perform. A shift of this kind is accompanied by many challenges. One will be to find employment for employees who are retrenched owing to the computerisation of their duties. It is obvious that jobs in the future will demand ICT skills. Learners who are taught ICT will be in a far better position to step into these jobs than those who do not have such education. The demand for professionals to work with computers is expected to accelerate this decade.

Workers who will be displaced are those doing duties that can readily be computerised. Mona (1999:13) explains that the benefits of ICT are not confined to learners, but extend to working adults, and qualifies this by saying, 'research has shown that employees who have basic computer skills, in addition to the requisite basic skills pertinent to their particular jobs, are more productive and have a longer career span'. Those who do not are quickly rendered functionally illiterate. How will organisations accomplish the necessary education to prepare these workers who may be rendered functionally illiterate? What about children who are still taught typing, which is 20 years behind the demands of the business world? These are questions with serious implications that must be addressed today so that we may handle the shift successfully and eradicate poverty.

Tulleken (2004:25) claimed that it is no secret that the government has been critical of current Internet governance structures because they were developed by and skewed in favour of those who initially employed these technologies. This negatively affected the people who are only now accessing these technologies.

Tulleken (2004:25) quoted the minister of communications, Dr Ivy Matsepe-Casaburri, who was addressing the ICANN (Internet Corporation for Assignment of Names and Numbers) meeting at Cape Town:

> We are of the firm view that we cannot continue on this path, thus perpetuating old relationships according to which developing countries, many of which were colonies of developed ones, were not allowed to build the necessary infrastructure for their own development. We believe that all should have a voice in the governance of an international network. We believe that legitimate governments, as the representatives of their country, should have an increased voice in the governance of the Internet.

The participation of the developing countries in addressing Internet inequalities and governance will not necessarily minimise the very important role that all other sectors have had in establishing and governing the Internet until recently. There has been a considerable change since the inception of ICANN as an initial step in that process.

The minister of communications stated (in Tulleken 2004:25):

> I am advised that ICANN itself has been through significant changes and continues to transform itself from a club of 'digital haves' to include the 'have-nots' and other marginalised groups. This can be seen in the increasing role of the Government Advisory Committee and the ccTLD [Country code top level domain] community in ICANN, the geographic diversity in its office bearers and the opening of an office in Europe and the planned establishment of an Africa office in the near future.

The minister said that two current ICANN initiatives were essentially close to her heart: the provisional recognition of the African Network Information Centre (AfriNIC) as a regional official Internet number registry, the last stage before full recognition; and the re-delegation of the South African ccTLD.za.

2.4.2 Meeting the requirements of legislation

The brain drain and the expectations of black people have placed increasing emphasis on the need for effective training. South African labour legislation is promoting training and affirmative action, which may lead to the hasty replacement of experienced white managers with less experienced black graduates (Employment Equity Act 1998). Lack of experience should be tackled by effective training. The government is aware of this, and thus the Employment Equity Act (1998), Skills Development Act (1998) and Skills Development Levies Act (1999) came into being. These acts compel employers to equip previously disadvantaged workers with skills that will make them well qualified and suitable for employment.

The legislation shifts the emphasis from certification to equipping people with the necessary skills for doing a particular job (Skills Development Act 1998). South Africa is lagging behind in terms of human resources development. The problem facing the corporate world is that executives are focusing only on filling vacancies. Most employers do not have a work development career plan and a suitable training model for their employees. CIE is one of the vehicles that could be used to roll out training.

How do employers carry out training effectively and efficiently? Companies cannot rely on traditional training methods because employees have to be taken off work during training, and productivity would be compromised. The trainee would move to the training venue and be away from the workplace. Cronjé and Baker (1999:43) observed that some organisations try to prevent the problem of interrupting work by placing the burden of on-the-job training of new appointees on experienced workers. This then increases the workload of the experienced workers, because the new employee is an additional responsibility. The implication is that productivity is reduced, because the trainee often interrupts the experienced worker's work. Organisations can therefore use online learning to handle on-the-job training.

The State of the Industry Report: Training and Human Resources Practices (ASTD 2003) shows an amazingly huge uptake in South Africa of online learning, including all computer-technology-assisted methods that allow interactivity and self-paced learning, with a number of organisations indicating that they are using or piloting online learning. According to this survey, development costs and lack of top management support are making it difficult for companies to adopt online learning. Other documents show that online learning in the corporate training environment was introduced around 1998 and is still in its early stages. Most organisations accommodate online learning in human resources departments, IT departments, learning centres, learning institutes, computer centres, corporate universities and virtual learning centres.

South African labour legislation was flawed prior to the new political dispensation in 1994. Swanepoel et al. (2003) reported that before 1994, some fourteen years after the great dawn of the 'new labour law' proclaimed by the eminent Wiehahn Report into industrial laws, every objective that laws were meant to serve remains unfulfilled. This assertion is supported by Rautenbach (Swanepoel et al. 2003):

But let us look around ourselves … South African workers are regarded by some as the most unproductive industrial workforce in the world; our strike rate compares favourably with the most strife-ridden industrial societies of the world; industrial violence and intimidation is commonplace; our unemployment is estimated to be at 40% of the active population.

2.4.3   Sector education and training authorities

One of the outcomes of legislation (Skills Development Act and Skills Development Levies Act) has been the introduction of sector education and training authorities (Setas). There are 25 Setas in South Africa whose purpose is to develop and improve the skills of workers. Setas are supposed to work out and execute sector skills plans, promote learnership, and hand out funds in their industrial sectors. However, the Setas are not performing well.

According to the *Business Report* of 28 January 2002 (as quoted in Swanepoel et al. 2003), most Setas are overwhelmed with teething woes. Problems with funding and logistics appear to have dominated the first year, according to at least 10 Seta annual reports tabled in parliament.

All the reports provide detailed descriptions of the steps taken to set up Setas to cover most economic sectors. Some, such as the banking, wholesale and retail and agricultural authorities, seemed to have had fewer teething problems than others. The Diplomacy, Intelligence, Defence and Trade Education Training Authority (DIDTETA), for example, did not have an industry training board under the old labour laws and had to begin from nothing.

On the other hand, (Swanepoel et al. 2003), several Setas pointed out delays in getting the South African Revenue Service (SARS), which brings together the levies, to transfer funds to them. This deferred payments to employers for training. Payments were sometimes wrongly calculated, while some employers had problems with filling in the required SARS forms. It seemed, however, that many of these troubles had been resolved over the year. Consequently, some Setas had substantial surpluses, while others had to be assisted by the DoL.

The DoL (2003) signed a memorandum of understanding with the Setas which would enable it to:
- Assess the individual performance of each Seta and its skills development contribution towards our National Skills Development Strategy
- Follow up on non-performing Setas
- Use the Quarterly National Skills Development Forum in which the National Skills Authority will follow up on Setas' performances as outlined in the quarterly reports they are obliged to submit to the department
- Empower Seta constituents to use these reports to monitor the performance of their Seta and take appropriate action. After all, it is their Seta and no official of the DoL serves on any of the Seta councils or boards.

According to the Skills Development Act (1998), Setas should promote learnership by finding workplaces where learners can do practicals, supporting people who create learning material, helping to wrap up learnership agreement material, and registering learnership agreements. However, some of the Setas are not doing this because of lack of capacity. The irony is that they have funds that are not being used. It was perhaps on these grounds that the minister of labour, Membathisi Mdladlana, called on Setas to spend the R2,8 billion surplus for skills development (DoL 2003). The minister also made a commitment to amending the Skills Development Act, so that he could deal with non-performing Setas. The South African Ministry of Labour has noted this and the necessary steps are being taken to address the problem.

2.4.4 Addressing the skilled workforce shortage

The shortage of skills in South Africa makes it difficult to close the knowledge gap. The 'Great Trek' of skilled workers exacerbates the shortage of skills. A large number of skilled people are leaving the country to pursue opportunities in the developed countries. Most of these professionals do not return. McLean (2002:22) reports that the brain drain continues to lead to the massive movement of intellectual capital from the country and our businesses. Transformation initiatives aimed at redressing the apartheid legacy and other imbalances in our economy will demand rapid knowledge transference to ensure that the full electronic support infrastructure is in place and that the success of affirmative appointments will improve. The long-term solution is to equip all employees with necessary skills. Ankiewicz (1995:248) supports this view:

> A further point in support of having a skilled workforce is that, as in the rest of the world, South Africa is in the midst of a recession. This has led to increasing unemployment and redundancy. Because of the short-sighted policies of the previous government, which discouraged the development of manpower skills in all sections of the population, there is a critical shortage of expertise. Because of the slow economic growth, most workers are being retrenched.

According to SABC News (2002) the National Labour, Economic and Development Institute (Naledi) produced a dark picture on the state of unemployment in South Africa. Ravi Naidoo, director at Naledi, reported that

> 100 000 full-time jobs were lost between June 2000 and June 2001, escalating the already huge unemployment rate in South Africa. Nevertheless, that is not the end of it, as more job losses are on the cards. About 16 000 jobs were reportedly lost in the mining sector. The manufacturing industry lost 50 000, the transport industry had to lose 13 000 jobs, while 19 000 jobs were lost in government.

Naidoo stated that the picture for 2002 is expected to be worse, adding that employment will not pick up in the formal sector, but it is likely to grow in the informal sector, which pays low wages. 'The sectors summit between labour/business and government should finalise agreements on how jobs could be created in all sectors of the economy. Leading sectors that could create jobs need to be identified soon' (Naidoo, SABC News 2002).

SABC News (2002) also reported:

> The South African Catering and Commercial Allied Workers Union (Saccawu) said
> that about 20 000 jobs are to be shed in the commercial and catering industries
> this year. Saccawu leaders met in Johannesburg last week to review their strat-
> egy on protecting jobs in their sector. Last year the union lost more than 20 000
> members because of retrenchments, and already a number of companies have
> indicated that they want to discuss retrenchments with the union.

2.4.5   The brain drain

Human resources development in South Africa is generally lagging behind the rest
of the world (see chapter 1). In fact, according to Cohen (1997), the previously advantaged (white population) are highly skilled. One would have thought that the
emigration of white professionals would provide opportunities for the previously
disadvantaged population groups (blacks, coloureds and Indians). However, in economic terms, Cohen (1996) reasons that

> such strategy is long-term and expensive and may result in short-term breakdowns
> in the delivery and quality of services. All patients would prefer to be wheeled out
> of an operating theatre alive, while all road users would expect that the bridge
> they drive over will remain upright whatever the ethnic origin of the surgeon or
> engineer.

The brain drain (increased mobility of skilled people) problem is exacerbated because even the previously disadvantaged people are leaving the country, especially
those in professions such as security, health and education (Ngubane 2002).

BBC News (2002) reported that thousands of highly skilled South Africans emigrate
in search of greener pastures overseas, 'draining the country of much-needed economic resources'. Most of them leave for English-speaking countries such as Australia, New Zealand, Canada, the US and the UK. These people are professionals with
rare skills that are in high demand in the West. Such skills include engineering, medicine, accounting and banking: 'most give fear of crime as the reason behind their
decision to go, but the Aids epidemic and unemployment are also cited in a recent
study carried out by the University of South Africa (Unisa, BMR 2004)'.

Several initiatives have been undertaken to minimise the brain drain and recall South
Africans in diaspora to contribute to the development of the country (Mbeki 2005;
Asmal 1999). In 2003, Dr Manto Tshabalala-Msimang signed a memorandum of un-
derstanding with the UK to decrease the poaching and recruitment of the South
African health professionals (Tshabalala-Msimang 2003).

But the brain drain can hardly be stopped by signing memoranda with other countries. Highly skilled labour is very mobile and this is a worldwide phenomenon. On
the other hand, it may not always be possible to quickly replace white professionals who have left the country with their black counterparts. This problem needs
long-term solutions. These include equipping more people with highly sought skills.
Computer-educated training may be one of the catalysts in addressing the brain
drain problem.

2.4.6 Difficulties in small, medium and micro enterprises

Since the dawn of the new democratic dispensation in South Africa there has been an increase in the establishment of small, medium and micro enterprises (SMMEs). The many reasons for this trend include the reduction of employment and eradication of poverty (Policy Coordination and Advisory Services 2006). It is perhaps on these grounds that the growth in the establishment of the SMMEs had been accompanied by the introduction of several programmes, initiatives and organisations to support them (Zake 2006). Bhagwana and Wall (2006) recorded that some of the SMMEs do not go beyond the start-up phase. This is owing to factors such as funding and managerial skills.

Several institutions of higher learning in South Africa offer courses, certificates, programmes and degrees in entrepreneurship. However, these academic offerings are not sufficient. The Policy Coordination and Advisory Services (2006) found that 'firstly, that a matric [Grade 12] qualification increases one's capacity to pursue entrepreneurial activities; and secondly, that tertiary qualification education increases the durability of entrepreneurial activity'. This is supported by table 2.1.

TABLE 2.1 NECESSITY AND OPPORTUNITY-MOTIVATED ENTREPRENEURIAL AC-
TIVITY AMONG YOUNG ADULTS BY EDUCATIONAL ATTAINMENT FOR
ALL DEVELOPING COUNTRIES IN THE GLOBAL ENTREPRENEURSHIP
MONITORING (GEM) (2003 SAMPLE)

	Not completed secondary schooling	Completed secondary schooling	Tertiary education
Probability of opportunity entrepreneurship (%)	5,8	7,5	12,6
Probability of necessity entrepreneurship (%)	6,1	3,8	3,5
Sum of opportunity and necessity entrepreneurship (%)	11,9	11,3	16,1
Ratio of opportunity to necessity entrepreneurship	0,95	1,97	3,60

Source: Adapted from GEM (2005), cited by The Policy Coordination and Advisory Services (2006)

One of the tenets of this book is that CIE has the ability to fast-track the roll-out of the entrepreneurship skills. Computers can provide training to many people in various places at once: something that a person cannot do practically.

2.4.7 Unemployment

Social factors such as high unemployment and diseases in South Africa necessitate the integration of CIE. There are vacant employment positions that cannot be filled because of the shortage of skilled labour. In other words, the majority of people who are not employed are the ones who do not have the necessary skills.

Unemployment is a serious social problem which is sometimes attributed to the lack of skills rather than the lack of jobs in the industry. According to the Global Poverty Research Group (2006), the rate of unemployment is one of the highest in the world since 2000 if one is using the broad definition. Kingdom and Knight (2005) warned that unemployment and underemployment in South Africa can become a 'threat to social and political stability' and thus the government and the commercial sector should place this problem at the forefront. In 2006 unemployment, according to Shezi (2006) of *Business Day*, was 'stubbornly' sitting at 26,7 per cent. The economy has to grow much faster in order to absorb people who need employment. If South Africa really needs to halve unemployment by 2014, economic growth should be above 6,5 per cent.

Unemployment will not be reduced by economic growth without developing the skills of the people. Many organisations have vacancies that cannot be filled owing to the shortage of skilled people (Blaine 2006). This assertion is supported by Le Roux (2006), who observed that the lack of 'skills in the construction sector is threatening the successful roll-out of government's public infrastructure development programme and projects planned for [South Africa's] hosting of the Soccer World Cup in 2010'. A study commissioned by Jipsa found that some South African graduates have qualifications that are irrelevant to the 'world of work' (Zake 2006). This makes the unemployment problem more complicated. Programmes such as Jipsa should therefore employ and promote the use of computers in human resources development.

Of course, nowadays there are young graduates who cannot find employment because they lack experience. The DoL is addressing this by introducing community service plans for these graduates. The government is also tackling Aids, which is killing some of our skilled labour. Another way of handling the implications of Aids is to train more people so that the economy is not adversely affected by the shortage of labour. CIE is also being used for this purpose. In one mining organisation, computer-integrated courses are offered to alert the workforce about Aids.

2.4.8 HIV/Aids pandemic

South Africa is currently experiencing one of the most ruthless HIV/Aids endemics in the world. AVERT (2006) reported that at the end of 2005

> ... there were five and a half million people living with HIV in South Africa, and almost 1000 Aids death occurring every day. A survey published in 2004 found that South Africans spent more time at funerals than they did having their hair cut, shopping or having barbecues. It also found that more than twice as many people had been to a funeral in the past month than had been to a wedding.

Although some of these assertions are questionable and debatable, HIV/Aids is a serious pandemic in South Africa.

The Bureau of Market Research (BMR, Unisa, 2004) predicted that HIV/Aids would have significant impact on household income and expenditure, 'as well as on national income in South Africa during the period 2004 to 20015'. In 2004 approximately one third of South African households were infected and/or affected by HIV/Aids, and that percentage was projected to increase enormously. According to the BMR (2004), the term 'infected' means circumstances where at least one family member is HIV-positive, while 'affected' entails that the household has to take care of at least one family member infected with HIV/Aids or an Aids orphan. This situation will have strong income and expenditure effects on millions of households in South Africa, with subsequent damaging impact on national income (Booysen, Van Rensburg, Bachmann, Engelbrecht and Steyn 2002).

HIV/Aids is having an impact on the economy of South Africa (Bollinger and Stover 1999). Aids is different from most other diseases because it attacks people who are predominantly at an economically active age and is usually fatal. Bollinger and Stover (1999) listed some of the major economic impacts of HIV/Aids:
- The loss of young adults in their most productive years will affect overall economic output.
- If Aids is more prevalent among the economic elite, then the impact may be much larger than the absolute number of Aids deaths indicates.
- The direct costs of Aids include expenditures for medical care, drugs and funeral expenses.
- Indirect costs include lost time owing to illness, recruitment and training costs to replace workers, and care of orphans.
- If costs are financed out of savings, then the reduction in investment could lead to a significant reduction in economic growth.

These sentiments are shared by Barks-Ruggles, Fantan, McPherson and Whiteside (2001), who observed that:

> Skilled personnel are lost and valuable labour time is consumed when workers become debilitated, and work schedules are disrupted when organisations replace workers and managers who are ill or have died.

The loss of capacity decreases economic growth. This premise has driven many large organisations to provide employment assistance services to employees (Pile 2004). Among other things, employers and organisations can employ CIE to provide some HIV/Aids-related lessons. On the other hand, CIE can assist in training more people so that South Africa can sustain its economic growth despite the pandemic.

2.4.9 High levels of poverty

The level of poverty is very high, especially in black communities (African black, coloureds and Indians). The government is addressing the question of poverty through legislation, programmes and projects. A large part of the national budget allocation goes to social development. One would conclude that the government is deliberately targeting black people for social and economic development and thus there is a policy on affirmative action. To a certain extent, this has succeeded in addressing poverty eradication among black people. However, the problem of poverty is becoming more complicated owing to the increase of white poverty.

2.4.10 The increasing poor white problem

According to SABC News (2004), the South African president, Thabo Mbeki, had noted the studies on the poor white problem: President Mbeki agreed to study statistics indicating a sharp rise in white poverty. He responded to data from the BMR (Unisa) that shows unemployment among white South Africans has increased by 200 per cent since 1994.

Referring to criticism that affirmative action might be to blame, Mbeki stated. 'If indeed there are consequences of government's actions which are resulting in greater impoverishment, clearly that is something we will have to look at … [and] see how we should respond.'

The study (The Bureau of Market Research (BMR), Unisa 2004) indicated that 400 000 whites (or about 10 per cent of the white population) lived below the poverty line, compared with none in 1990, Willie Spies, a Freedom Front Plus lawmaker, said. South Africa's overall unemployment rate is estimated at between 28 and 40 per cent, and is most severe among poor rural blacks. The UN 2003 South Africa Human Development Report released this year said 21.9 million people, or 48.5 per cent of the total population, lived below the poverty line. Overall, South Africa remained one of the world's most inequitable countries, with average annual spending in white households being six times that of their black counterparts, and with income disparities widening.

In fact, President Mbeki has observed the white poverty problem (Tsedu, Sikhakhane and Jeffreys 2004):

> Personally, I had not realised the extent of the poverty among the poor sections of the white community … It has been quite disturbing where, for example, in Cape Town young white women … said: 'Minister, we have to work as prostitutes because we can't maintain ourselves, we can't maintain our children, but the police harass us in the streets. Can't you please talk to the police to just leave us alone, for there's no other way to make a living.' They [young whites] also expressed surprise that the government actually cares about their own conditions and their future.

CIE should be employed as another tool for eradicating poverty in South Africa. Addressing poverty through government social grants may not be sustainable. The author doubts whether the basic income grant, which has been sought by Cosatu and other South African organisations, will succeed in lowering the effect of poverty in the long term. The Basic Income Grant Coalition was established in mid-2001 to develop a common stance among supporters of a universal income support grant and to mobilise popular support for the introduction of the grant.

Nevertheless, President Mbeki's cabinet was deeply divided regarding the basic income grant, which would help about 23 million people who live in excessive poverty. Dr Zola Skweyiya, minister of social development, was the only cabinet minister who publicly supported the basic income grant. This form of assistance would be targeted at poor people who qualified for the 'existing grants such as the social, disability and child support grants or an old-age pension' (Ensor 2006). According to Ensor (2006), until now, the South African government, especially Minister of Finance Trevor Manuel, has warned against huge spending commitments from which it will be impossible for the state to disentangle itself, should the country experience economic downturn. The long-term solution for poverty eradication is skills development. Black economic empowerment (BEE) initiatives will be successful only if they are driven by a skilled workforce and skilled entrepreneurs.

CHAPTER 3

MACRO-ORGANISATIONAL ISSUES

CONTENTS

3.1 INTRODUCTION

When an organisation introduces CIE, it is advisable to recognise and address organisational issues. The implementation of CIE should not be done in isolation from other organisational functions. Organisational goals and plans should play a prominent role in the integration of computers in human resources development. Top management and human resources development managers should have a broad understanding of the potential of CIE in the transformation of human resources development. But top management may not provide the necessary resources unless they are convinced of the value that CIE will add in the organisation. A lack of human and infrastructural resources may impede the implementation of CIE.

3.2 ORGANISATIONAL ISSUES

Before CIE is implemented in the corporate training environment, several organisational issues should be attended to:

3.2.1   Top management support

It is very important to obtain the support of top management when implementing online learning (Tetiwat and Igbaria 2000). A person at the level of director should sponsor the online learning initiative. He/she should sell the strategic idea of CIE to the top management and board of the organisation. The implementation of CIE requires considerable resources, so board approval should be sought (Johnson 2002). Palloff and Pratt (2001) suggest that a committee that includes managers from all business units should initiate CIE.

In South Africa, most organisations that are implementing online learning have the support of top management. Some, however, are still struggling to obtain this support. This may be because the team responsible for online learning does not have an online learning strategy or a business case. According to Mayberry (2003), a business case reflects gap analysis. It shows the organisation's current status versus the prospective status and how the prospective status will be achieved.

Documents that were collected from one of the South African largest banks revealed how the managing director insisted on seeing the benefits of e-learning before he would allocate funds and resources to a full-blown integration of computers for education, training and development.

3.2.2 Building a business case

According to Mayberry (2001), a business case is a tool that reflects a gap analysis. The business case describes the enterprise's current situation versus the desired status. It should also show how the desired status would be achieved (Willis and Kelly 2002:38). A well-developed business case discusses how planning and decision making regarding procurement, outsourcing and integration strategies will be conducted (Metzler 1999:27).

Although there are many advantages to building a business case for CIE, Mayberry (2001) notes that most human resources and training professionals do not know how to create it.

Most of the organisations selected for the study did not have a written business case or strategy for CIE. Only a few were in the process of developing one. Most organisations have implemented CIE without a business case, even though top management is reluctant to support CIE initiatives in its absence.

The researcher (author) found that there was generally a lack of essential documents on the integration of CIE in the corporate environment. Some organisations regarded four-page PowerPoint documents as their CIE strategy or policy. Most organisations were therefore implementing CIE without business cases, strategies and policies. It is perhaps on these grounds that in some of these organisations top management was reluctant to provide resources for CIE endeavours. However, most executives have supported CIE in the absence of these crucial documents.

3.2.3 Involvement

People usually support what they helped to create (St Clair 2002:1). Involvement in CIE should not be confined to managers. Every employee, unit and structure within the organisation should share ownership of the CIE initiative. In a unionised country such as South Africa, unions could also be involved (Boxer and Johnson 2002:37).

Employees should be provided with diverse learning programs and styles (St Clair 2002:2). CIE should be made available to each unit in the organisation. CIE coordinators should obtain potential course requirements from business units. In some instances, outside content providers could be consulted. The content provided by these providers should be customised to suit local training needs (Boxer and Johnson 2002:41).

Line managers should be accountable for the performance of their staff in CIE. Managers should be role models by participating in and completing CIE courses (Anon 2001c:56). This will encourage workers to complete their online courses according to the schedule.

CIE learning implementers, according to Boxer and Johnson (2002:37), should resist the temptation to implement a fully fledged CIE project quickly. A fully fledged project could be preceded by a pilot project. Feedback should be gathered from the pilot participants, adjustments made, and the pilot experience built on (Baxter 2001:14).

Most of the organisations that participated in the study are implementing online learning gradually. They usually start by doing pilot projects. In some instances, pilot projects are done before the business case is finalised, and the data gathered during the pilot study is integrated into the business case. This information enables the organisation to identify aspects that are appropriate and those that should be modified or dropped. The pilot project also guides the organisation when it comes to choosing the learning management system and business partners. One major telecommunications company initiated CIE in 1998, and its virtual campus accommodates more than 20 000 registered learners (employees). However, most organisations started only in 2000, and the majority are still in the planning stages.

All the organisations that participated in the study indicated that they enjoy a great deal of support from business unit managers and employees. However, they still need to 'sell' CIE and introduce a learning culture in the workplace. In some organisations, business unit managers adopt the role of promoters, encouraging their subordinates to participate in CIE courses and assessing their progress. This may account for the high competition and pass rate. In some instances, managers allow employees to access the virtual classroom during working hours, as long as it does not affect their job performance. Employees who participate in online learning are expected to carry some responsibility, especially when it comes to time management.

In South Africa, managers provide learning coordinators with a list of learning needs, which are then converted into learning content/courses. The conversion process involves managers, instructional designers and trainers. In some instances managers serve as subject matter experts, and employees/learners are involved during the pilot project. It is advisable for instructional designers to work very closely with subject matter experts to create a detailed instructional strategy that includes learning activities, information resources and evaluation for online learning materials. This process should be a systematic and reflective one, which takes due account of principles of learning and training. The ADDIE model (analysis, design, development, implementation and evaluation) would be a useful tool in this context.

Other courses are very general in nature and business managers do not necessarily suggest them. These include computer skills and management and leadership courses, and are mostly bought off the shelf. The problem with off-the-shelf courses is that they do not necessarily address the specific and unique needs of a particular organisation. In addition, they are mostly designed and developed in other countries, so they do not always meet the requirements of South African legislation, such as outcomes-based education and training, SAQA and the NQF. This is not to say that these courses are of poor quality; most comply with quality standards of international and business associations.

3.2.4 Online learning as a focal point

Organisations should use their corporate Intranets as a portal for CIE (Slabbert and Fresen 2002:2). This will make the virtual campus visible because employees usually visit the Intranet. Boxer and Johnson (2002:40) argue that CIE should become a focal point if it offers courses that add value to the daily activities of the workers.

When employees use their desktops to participate in CIE, a 'do not disturb' sign could put on their desks so that they are not constantly distracted (Anon. 2001c:56). Managers should respect the staff's CIE time. Alternatively, employees could leave their desks and learn in the CIE centre (Boxer and Johnson 2002:42).

Online learning should enable employees to take responsibility for their personal professional development, and courses should be linked to business strategic priorities (Tucker 1997:34). Online learning should provide line and functional managers with resources to improve employees' performance and professional development (Ludlow, Foshay, Brannan, Duff and Dennison 2002:37; Conee, Shackelford, Boxer, Johnson and Weaver 2002:72).

CIE courses should be promoted through articles, pamphlets and advertisements on employees' desks, in internal publications, and in posters displayed in the canteen and at other gatherings within the company (Boxer and Johnson 2002:41). This will assist in making online learning a focal point in the organisation.

Many organisations house learning centres, learning institutes and computer centres that allow employees access to computers and provide an environment in which employees may learn without being disturbed. In one mining company, some employees work in shifts and use some of their free time to access the learning courses. A learning environment such as this is advantageous because learners are able to interact with trainers who can offer assistance if necessary.

All the organisations that participated in this study believe that the focus should not be on CIE only, but on learning as a whole. CIE is therefore integrated with traditional learning methods. Consequently, participants spoke extensively about blended learning during the interviews. The various centres, institutes and universities also adopt a blended approach. In the corporate training environment, blended learning means:

Combining Internet technology with the virtual classroom, self-paced learning, collaborative learning and online videos
• Combining training methods
• Combining instructional technology with face-to-face trainer-led training
• Combining learning with real job tasks

The organisations that took part in the study are apparently aware of the danger of being swept away by the excitement of the Internet that has led many organisations to make bad decisions. They are using the Internet to complement their traditional training methods, and CIE is not separated from the established learning operations. Organisations are therefore using the Internet to build on proven principles of effective training.

3.2.5 Support and training for facilitators and learners

CIE facilitators should be supported when they deal with course design and development, and learners' issues (Palloff and Pratt 2001:154). A facilitator who is dealing with an uncooperative learner needs to know the parameters of his/her efforts in order to deal with the problem constructively (Drysdale 2002:6). There should be policies and guidelines that facilitators can use if they encounter difficult situations (Palloff and Pratt 2001:154). Similarly, the organisation should ensure that learners are assisted swiftly and fairly when they encounter technical or educational problems (Slabbert and Fresen 2002:15). If neither facilitators nor learners feel a sense of organisational support for actions they want to take in the virtual classroom, the consequences could be disastrous, and some learners may drop out (Palloff and Pratt 2001:154). Campbell, Sawert and McPhee (2002:3) claim that providing CIE goes beyond creating a virtual classroom and hoping for the best.

Campbell et al. (2002) report that learning facilitators and learners should be trained to use the learning management system applied in their organisation. Training should enable participants to understand the need for change and to maximise online interactivity (Van der Westhuizen, Stoltenkamp and Lautenbach 2002). In some instances they should be trained in computer basics (Nxasana 2002). Facilitators should be trained in designing, developing and customising CIE materials (Lautenbach and Van der Westhuizen 2002). They should be trained in how to deal with learners and their problems in the virtual classroom, and how to accommodate learners' individual styles and group dynamics (Palloff and Pratt 2001).

Where a learning management system is not used, learning facilitators should have a broad knowledge of web-based technologies and should acquire website development skills (Tetiwat and Igbaria 2000; Asmal 2002). Walker (1999) reports that 70 per cent of organisations implementing online learning worldwide are spending about 70 per cent of their computer-integrated budget on training facilitators and users. This shows that training is one of the success factors of online learning. Tetiwat and Igbaria (2000) suggest that basic computer skills should include setting up and operating computer hardware, software and the learning management system. This will enable facilitators and learners to be comfortable and efficient in the CIE environment. According to Campbell et al. (2002), some organisations employ a support team that helps users. This team needs to be highly trained in order to train the users effectively (Place, Stephens and Cummingham 1999). The organisations that participated in the study provide professional and technical support to learners and learning facilitators. Vendor consultants work full time on the client's premises each day to help learners and facilitators. However, most organisations prefer their e-learning staff to be empowered to render the support service themselves.

One form of support that online learning coordinators need relates to the production of the learning content. They do not have this skill themselves, which is why content is bought off the shelf. Some organisations have outsourced this responsibility to their vendors.

Learning facilitators and learners in the organisations that participated in the study receive orientation before they engage in online learning activities. This enables them to use the learning management system's tools. However, in some organisations, learners are expected to be competent in using the keyboard and mouse, managing files and using the word processor, databases, e-mail and the Internet. It will be shown in the next sections that training to learners and facilitators was not always adequate (Walker 1999).

3.2.6 Curriculum issues

Zafeiriou, Nunes and Ford (2001) suggest that CIE curriculum design should ensure that learning activities are integrated into real-world contexts. Learning activities should also provide authentic and multiple perspectives on the subject being dealt with (Cloete and Miller 2002).

Learning facilitators should provide coaching, guidance and examples to CIE learners (Kruse and Keil 2000:126). Subject matter experts may be sought to provide different perspectives of the subject matter (Morrison 2003:387). This, according to Zafeiriou et al. (2001:97), will provide learners with skills such as negotiation of meaning, lifelong learning, reflective analysis and metacognition.

The organisations that participated in the study try to ensure that curriculum design and learning activities are integrated into employees' real working activities. However, these efforts are undermined because most courses are general in nature, are obtained from other countries, and are bought off the shelf. The establishment of learning centres, institutions and corporate universities would solve this problem. Some organisations have fully fledged teams that deal with content design and development. In some instances, these centres are independent of the organisation's human resources departments and serve the organisation's clients and other business partners.

3.3 HARDWARE AND SOFTWARE ISSUES

Managers responsible for the implementation of CIE should consider hardware and software issues when implementing CIE.

3.3.1 Hardware issues

When using the Internet for online learning, facilitators and learners may need a personal computer (with sufficient memory of RAM and hard disk), modem, printer and a CD-ROM drive (Forsyth 1998:143). The hardware equipment could be very costly to an organisation that has many employees. Although the employer may provide learners with hardware equipment at their workplace, learners who want to learn after working hours, on holidays and at weekends may feel compelled to purchase the expensive equipment (Palloff and Pratt 2001:57).

Forsyth (1998:143) advises e-learning facilitators and organisations to bear in mind that technology and software are evolving, and users may be required to upgrade their equipment in the future. This may lead to educational and computer cost dilemmas where the cost of the technology delivering the course outstrips the user's ability to finance these expenses in order to complete the course. Of course, this cannot be experienced in a short course (Hoole and Hoole 2000:33). The organisation should therefore take changes in technology into consideration when contemplating CIE.

Educational considerations should influence the selection and acquisition of computers (Palloff and Pratt 2001:57). Buying computers and software packages without a plan of how they will be used for online learning purposes is short-sighted. Gehring (2002:28) thinks that understanding issues around instructional technology is essential in the organisation to avoid pitfalls associated with the failure of integrating CIE.

Palloff and Pratt (2001:57) list some questions to consider before technology equipment can be acquired:
- How does/do the organisation or individuals envision using technology? Will it be regarded as a support to the face-to-face classroom or will it be used to deliver classes and programs or both?
- How will technology meet instructional needs in a particular organisation?
- What do people see as the implications of technology in their organisations?
- How extensive do the organisations expect the use of technology to be? Will it be phased in or will the organisation move directly to offering full programs that require immediate implementation?

In most of the organisations that participated in the study, educational considerations influenced the selection and acquisition of computers. Learning and human resources development managers in these organisations consider understanding issues related to instructional technology to be essential within the organisation so that pitfalls can be avoided.

The organisations investigated do not spend so much time focusing on technology that technology dominates the training and learning activities and the subject being delivered is ultimately lost.

3.3.2 Software issues

Widespread and reliable communication tools, software and services that support training, learning and collaborative work have accompanied the emergence of the World Wide Web (WWW) (Klobas and Renzi 2000:43). Effective software and services provide efficient opportunities for enhancing CIE and learning strategies and activities (Palloff and Pratt 2001:157).

According to Taylor (2002:12), one of the things that makes CIE attractive in the corporate training environment is the software option or learning content management system (LCMS). LCMSs make the CIE course easy to access and manage (Arnone 2002a:33). The systems enable learning facilitators to manage courses by facilitating, assessing and discussing learning matter online (Richards 2002:1). Taylor (2002:11) argues that no matter what kind of software platform the organisation has employed, the most important thing is to ensure that the quality of online learning is at the same level as traditional learning and training.

Arnone (2002a:34) claims that most organisations prefer to use the software that enables them to break the learning content into manageable chunks. Gehring (2002:28) advises that the chunking of content should be in line with the instructional strategy. Arnone (2002a:34) declares that the best way of breaking down content is to use content objects/tools within the learning management system. These allow learners to pick up and choose the content they want to utilise and enable facilitators to rearrange content easily to accommodate individual learning styles. Assessment tools can track learners' progress through the learning/course management system.

One of the challenges of dealing with learning management systems is to develop selection criteria (Klobas and Renzi 2000:44). Table 3.1 reflects criteria that learning facilitators could use as a model for selecting a learning/course management system. It also reflects CIE tools.

TABLE 3.1 SELECTION CRITERIA FOR EDUCATIONAL SOFTWARE (LEARNING MANAGEMENT SYSTEMS)

Educational strategy	Characteristics of strategy	Category of learning management system
Lecture or presentation	Facilitates and presents learning materials to virtual classroom	Readings or presentations converted to web page index of downloadable material (text, tables and presentations) or audio/video material live or recorded and distributed via streaming technology
Workshop or laboratory	Learners complete set tasks designed to develop their skills. Often live or recorded demonstrations presented or prepared by a facilitator are included	Activities prepared using learning management system, including multimedia technologies, are made available to learners from a web page
Self-guided learning	Learners work individually (often isolated geographically) to complete assigned readings and exercises	Readings, references and activities, prepared using learning management system or WWW technology
Seminar or tutorial	Learners, working in relatively small groups, discuss set topics, cases or readings, with the facilitator's guidance	Discussion or conferencing software
Consultation	Learners (individually or in small groups) meet with the facilitator to obtain answers or guidance on topics	E-mail, chat, audio and video conferencing

Collaborative learning	Learners work together; they learn through collaboration with one another rather than from material delivered by the facilitator	Discussion or conferencing services, e-mail, chat, audio/ video conferencing, specific tools for community building and collaborative work

Source: Adapted from Klobas and Renzi (2000)

The learning management system selected should serve not only the educational considerations, but also those related to task, training administration and technology dimensions (Klobas and Renzi 2000:44). In other words, an efficient and effective learning management system should have all the foregoing dimensions.

3.4 PERCEPTIONS OF COMPUTER-INTEGRATED LEARNING IN THE CORPORATE TRAINING ENVIRONMENT

3.4.1 Managers and facilitators

During the fieldwork, it was found that CIE managers and facilitators have certain perceptions regarding the use of computers for educational, training and developmental purposes. It is therefore important to know how role players perceive CIE.

a) A convenient way of learning

The managers and facilitators perceived online learning as convenient: 'The courses are online, self-paced and they are available at the desks.' Employees (learners) do not have to leave their offices 'and stay away a week from work or two days'. Employees sit and learn in their workplace 'at any time'. The managers and facilitators in this study therefore saw online learning as a 'big opportunity for the staff to develop themselves' in a convenient way. They were very impressed with the individualisation associated with CIE because employees could learn 'at their own pace and finished training' within the stipulated time. This is enhanced because 'assessment is automated and results are also automatic and auditable'. In some instances employees could learn at their own homes. CIE allowed them to learn 'anywhere, at any time and it is more efficient, structured, self-paced, immediate' and offered 'individual feedback'.

b) A simpler way of training employees in different venues

Managers and facilitators perceived CIE as a simpler way of offering training to employees based in different countries: 'The online learning pilot project which was introduced here in South Africa was extended to workers in the UK, Australia, Canada, Brazil and Botswana.' However, even within South Africa, employees of organisations are scattered around the country: 'If you've got more than 2 500 employees, you are geographically dispersed – you have to find a way to deliver the training' in a more convenient and consistent way. CIE enables a mining conglomerate to deliver training in its mines 'in the southern part of Namibia, near Luderitz and in the Northern Cape; in the North West Province – on the Zimbabwean, Mozambican border. Then Richards Bay, in Ellisras, Tabazimbi' and somewhere in the Gauteng area. CIE is very useful in delivering training to the South African banking sector: '[I]f you look at our rural infrastructure, our rural branch network for example, you've got branches situated here, others situated 50, 100, 500 km away and the campuses where we conduct our learning are also far situated.' With online learning they are able to breach the gap. They were able to deliver learning to employees who were really far from where learning usually took place.

c) A cost-effective way of training

Managers and facilitators who participated in this study perceived CIE as a cost-effective way of training because 'travel, training and administrative costs' were reduced. The costs of traditional learning are usually very high and 'so we had to find a way to deliver the training in a more cost-effective way'. The managers and facilitators noted that if employees left the workplace and went to the central training venue, 'they claim money, travel allowance, extra'. Employees would also have to have 'meals there and everything' and thus the organisation 'saves on that side as well' if online learning was integrated within the company. The development of paper-based learning materials is very expensive, and CIE could offer learning materials in a cost-effective way. In fact, some managers and facilitators claimed that 'it's a lot cheaper than face-to-face classroom learning too' and the organisation can 'attain a good return on investment' quickly. Costs of traditional learning were exacerbated because 'in most instances employees have to be flown in from around the country, booked in at hotels, fed and transported – all at huge cost' and with great disruption to their work schedule.

d) The value of learning

Most managers and facilitators were more interested in the value accumulated from CIE activities than the return on investment. Although organisations initially employed online learning with the 'perception that it could cut costs', they had now moved away from costs to the quality and value offered by CIE. What was attained from online learning could be gauged by conducting pre- and post-online learning assessments: 'Learners were given a pre-assessment, then worked through the e-course on products and were then given a post-assessment to complete.' The assessment results were impressive. Managers believed that they needed to prove the value of CIE, but they did not necessarily 'have to prove the high return on investment'. In fact, some thought that 'ROI [return on investment] does not exist. It's nonsense! We are trying to measure something in a new paradigm in an old paradigm.' This view is supported by Morrison (2003:56).

e) Other factors

Most of the managers and experts perceived the integration of CIE to be a result of certain factors such as the 'availability of computers and network infrastructure, growing number of computer applications, increased complexity of work, rising training costs, demand for just-in-time training, demand for workplace training' and legislation. This legislation compels South African organisations to train their employees. Organisations should have special development programmes for the previously disadvantaged. Training can also be achieved by interacting with the Setas. Another reason for the integration of online learning in the corporate training environment for human resources development 'is statutory requirements'. In South Africa, it has become 'a legal requirement for companies since year 2000 to pay 1% of the payroll to the skills development fund'. The use of CIE for training enables organisations to reclaim the money they have paid because of the requirements of the Skills Development Levies Act.

3.4.2 Learners

a) An important method of training

Learners perceived online learning as an important method for human resources development in the corporate training environment: 'From my side I learned that it's very important.' Employees found online courses essential for their skills development: 'The courses that they put on the system are very, very important.' Learners also found that 'there's a lot of courses that you can choose from' and 'there's a lot of information inside these courses; maybe even more than if you had gone to the classroom'. In face-to-face training 'the lecturer can't always give you all the information' that can be obtained in CIE courses.

b) A new and intimidating delivery mode

Some learners perceived CIE as a new training delivery mode: 'That is something very new because we are very used to conventional classroom training.' Other learners were intimidated by 'new technology; most people are very scared'. This could be attributed to the age of the learners. Older employees were not really comfortable with CIE: 'The majority of our learners are not very young.'

3.5 DEALING WITH RESISTANCE

Resistance sometimes confronts the use of computers for education, training and development. During the interviews, the researcher noted that there were several reasons for resistance to the implementation of CIE. Some employees felt that they lacked the skills and behaviours required in the CIE environment. Organisations solved this problem by introducing pre-training, which acquainted employees with computer literacy, general literacy, Internet skills and the use of a learning management system. Some facilitators who had been responsible for traditional face-to-face training resisted CIE and sabotaged the implementation of the CIE strategy, fearing that they (the facilitators) would become redundant. Organisations solved this problem by offering training to the facilitators who were responsible for CIE.

Organisations also assured stakeholders that CIE would not replace traditional learning, but that the two would be integrated. During the interviews, the concept of blended learning was mentioned several times. In South African organisations, blended learning refers to combining Internet technology with the virtual classroom, self-paced learning, collaborative learning and online videos; combining training methods; combining instructional technology with face-to-face trainer-led training; and combining learning with real job tasks.

Some people did not understand the benefits and other implications of implementing CIE, and so resisted it. In some instances, this resistance was dealt with through the development of business cases for CIE, preceded in some instances by a pilot project. However, most of the organisations implemented CIE without a business case, and the resistance continued.

In some instances people resisted the implementation of CIE because they thought they would be at a disadvantage because of the lack of resources. This happened mainly in branches in underdeveloped areas where employees lacked access to computers and/or the Internet. This problem was solved through the introduction of computer centres, corporate learning institutes, virtual learning centres, corporate campuses and one corporate university.

Inter-organisational agreements were sometimes a cause of resistance. Labour contracts and agreements with unions affected the implementation of CIE positively and negatively. Although the unions support training and development initiatives (and CIE in particular), the question of when CIE should be carried out remains unresolved. Questions include when employees should participate in e-learning activities, and whether managers will allow employees to use work time for CIE. Some of the organisations dealt with change management by assessing the business environment and determining their enterprise's performance gaps. This made it possible to implement CIE to close the performance gap.

3.6 BENEFITS OF COMPUTER-INTEGRATED LEARNING

When integrating CIE into the corporate training environment, the organisation should appreciate that CIE has benefits and limitations.

3.6.1 A consistent message

One of the most important benefits that persuaded corporate learning facilitators to implement CIE is that it offers a consistent message (Tucker 1997:24). That is, if you offer a lesson through CIE to another group of learners, it will offer the same message, unlike lessons presented by face-to-face learning facilitators, which will be affected by several factors. For example, face-to-face learning facilitators have their own moods, learner participation may divert them, and they cannot tailor their lesson to individual learners. Some learners are slow, and others have their 'off days'. In fact, a person cannot present the same lesson consistently to different groups, and thus the integrity of the message is questionable (Hannafin and Peck 1988). Nonetheless, the need for a consistent message depends on training needs and circumstances.

Tucker (1997:24) provides an example of a multinational accounting company that used interactive CIE to train its accountants in 40 countries worldwide. By using CIE, the company was sure that its accountants in the UK and the rest of the world would receive the same quality of training, and therefore apply the same accountancy practices throughout its offices.

3.6.2 Employees learn more

Learners learn better through online learning because it creates the same level of playing field for all learners. Learners who have poor confidence or some disability may find it difficult to participate in a face-to-face class and become withdrawn (Goral 2001:54). However, in CIE all learners are equal. CIE allows learners to learn more because of the following advantages:

- CIE allows training to be offered to employees or small groups across a broad geographical area or a number of isolated sites, even the most remote areas.
- CIE eliminates time and costs associated with travelling to attend in-service training workshops and courses for learning facilitators and participants.
- CIE interaction formats provide a mechanism to support ongoing staff development programmes using coaching and mentoring over weeks or even months, with opportunities for implementing changes and feedback on new practice.
- CIE can support the formation of communities of practice around a common theme by facilitating interactions and sharing of ideas that eliminate the boundaries of time and distance.
- CIE helps practitioners to acquire advanced technology skills that help them make better use of the Internet to locate resources and network with colleagues (Ludlow et al. 2002:41; Sinclair 2001:32).

3.6.3 Convenient for workers

CIE allows learners to learn at their own pace (Tucker 1997:30). In face-to-face training, the pace of the training session accommodates neither the slow nor the gifted learners (Tiffin and Rajasingham 1995:143). Some learners find the pace too fast, whereas others find it too slow, and thus learning becomes difficult for everyone. Learners who find the pace too slow become bored and their attention is distracted. Learners who perceive it as too fast feel under pressure, and switch off (Tucker 1997:30). The perceived slowness or fastness affects learners' contribution and involvement negatively (Warbington 2001:23).

Adult learners are not usually enthusiastic about attending a residential course at the expense of family and other obligations (Taylor 2002:11). They tend to become distressed at leaving their families. An illness in the family can exacerbate lack of interest in attending a residential course. However, with online learning the learner can learn at home while being close to the family (Tucker 1997:31). CIE will allow the individual learner to omit sections that he/she is already familiar with or supplement a specific lesson with some relevant material (Taylor 2002:15).

In most instances employees need just-in-time training (Cronjé and Baker 1999:44). This kind of learning enables them to use knowledge and skills acquired in their specific tasks and to add value in their jobs by transferring knowledge (Tiffin and Rajasingham 1995:144). CIE can allow corporate learning facilitators to offer a course that would assist employees to perform certain tasks immediately.

Cronjé and Baker (1999:44) believe that an electronic performance support system could meet employees' needs. In addition, applying acquired knowledge and skills in the workplace can enable the employees to retain knowledge and skills for a longer period (Fuller, Awyzio and McFarlane 2002:3; Tucker 1997:25).

3.6.4 Easily retrievable materials

Forsyth (1998:13) observes that there is general excitement in the learning community about the links in the virtual classroom that connect the learner to vast learning resources. Stephens (2002:1) notes that in most instances the CIE facilitator posts notes, learning materials and assignments in the virtual classroom prior to the training. This enables learners to prepare before the training. The handouts are characterised by diagrams and pictures that illustrate aspects of the lesson. Digital movie clips may be used to demonstrate procedures and directions (Stephens 2002:1; Poole 1998:128). In the electronic laboratory and workshops, learners can easily retrieve the relevant learning materials that will enable them to conduct certain experiments (Tucker 1997:26). On the other hand, learners can attach notes to their e-mails so that they can use them at home. Stephens (2002:2) suggests that simulations could be used virtually to supplement the traditional laboratory.

Tucker (1997:26) notes that virtual classrooms are usually available at the workplace (in organisations that have e-learning), and thus employees can easily retrieve the section that they want. However, in a traditional learning environment, learning materials are usually kept in the physical library, and it is not easy for working people to frequent the library.

3.6.5 Cost-effective

Tucker (1997:27) and Arnone (2002b:27) claim that online learning is usually more cost-effective than traditional learning. However, authors such as McCormack and Jones (1997:22) object to this claim. (Their argument will be reflected in the paragraphs that deal with myths about CIE.)

In big corporations the set-up cost of CIE is distributed over a large number of employees, making it cost-effective (Tucker 1997:27). The cost-effectiveness is particularly advantageous in training that involves people in different parts of the country or world (Capper 2001:238). Tucker (1997:27) argues that the cost-effectiveness of traditional training is difficult to gauge because it is usually done in-house.

Tucker (1997:28) lists areas in which savings can be made by using CIE:
• Training time is reduced, because people learn more quickly.
• Time away from the job is reduced because travel is not involved.

- Learning may take place outside working hours.
- Learning may take place in quiet periods.
- Travel and accommodation for traditional courses can amount to as much as 20 per cent of the total training costs.
- Time away from the job is a significant cost, particularly where it is necessary to employ agency staff to cover for members of staff being trained. In other instances, staff members may have to work overtime to cover for their absent colleagues.
- Lost business opportunities can be significant when looking at the cost of traditional training, although it is not possible to quantify business lost owing to staff being away on training courses.

These areas of cost savings motivate organisations to use CIE. The next section will deal with the limitations of CIE in the corporate training environment.

3.7 LIMITATIONS OF COMPUTER-INTEGRATED LEARNING

3.7.1 Learners may feel lonely and frustrated

Although CIE enables learners to learn at any time, anywhere and at their own pace, learners miss the physical interaction with other learners and learning facilitators. McLester (2002a:25) refers to physical interaction as the 'human factor', which has a prominent role in the learning process. When conducting research on online learning, McLester (2002b:4) realised that even learning facilitators miss the 'collegial relationships and in-process rapport' with their learners. Learning facilitators indicate that physical interaction gives them a lot of the intangible rewards of facilitating learning.

Learners also miss the daily back-and-forth with other learning facilitators and connections with their peers in the training lounge (Leyell 2002:1). Employees who are learning online miss the youthful energy they had during their face-to-face schooling days in the hallway and classes (McLester 2002b:4). McLester also notes that some corporations acknowledge the impact of the human factor and therefore ensure the presence of a local learning facilitator in each branch who mentors employees and ensures that they are progressing positively. Mentors who have specific interacting time with learners create the most effective online learning environment. Peer interaction is also essential. Learning is social, and sometimes people need 'heart to heart' interaction. Employees can learn from each other through body language and voice tone. They learn more in the reflection of other employees than by seeing their reflections of themselves. CIE is like looking into the mirror as introspection, but if it is overdone it could lead to 'solipsism and self-absorption'.

According to Lautenbach (2000:25), the unavailability of face-to-face learning makes some learners frustrated and uneasy. This is exacerbated because the 'lack of facial expression, body language and tone of voice leads to misinterpretations, unintended insults and ambiguities, which may detract from the aims of the course' (Hall 1997, in Lautenbach 2000:25). The lack of competition and pressure associated with social interaction in traditional learning contribute to loneliness and isolation in the CIE environment (McCartan 2002, as cited by Lautenbach 2000:25).

3.7.2 Lack of familiarity with technology

One of the requisites of computer-integrated learning is familiarity with computers (Zafeiriou et al. 2001:87). Lack of familiarity with technology can hamper CIE and affect the learner's confidence negatively (Lautenbach 2000:62). Learners should be familiar with the hardware and software. Learners who have not used computers before may even be scared to press the buttons in case they break them (Zafeiriou et al. 2001:87). Familiarity with technology can make a huge difference because the learner should feel comfortable when using computers (Walker 1999:274).

Technical problems can frustrate the learner, and in most instances learners do not meet the technical requirements of the course (Lautenbach 2000:26). Some learners think that they will not be able to get assistance to deal with computer glitches and the complexity of the software (Zafeiriou 2001:87). Lack of familiarity with technology also affects online group participation and results in uncertainty (Lautenbach 2000:26). As a result, some CIE group members may feel isolated and become reluctant to participate in classrooms and discussion boards (Zafeiriou 2001:88).

According to Walker (1999:275), computer and Internet skills are basic conditions for participating in computer-integrated learning. Learners with limited computer and Internet skills will not be able to participate fully and thus will be disadvantaged (Lautenbach 2000:26). In most instances, learners who are more familiar with technological technicalities dominate online activities, and this is not always fruitful in the learning environment (Zafeiriou et al. 2001:89). Lautenbach (2000:26) observes that older learners with little exposure to technology may lack online learning collaborative skills and may not learn much from their classmates.

3.7.3 Poor interactivity

Most online courses are sequentially structured and have series of screens that contain content and graphics (Phillips 1997:26). The learner is expected to move through the screens consecutively. But the learner may do so without being seriously engaged in the learning process (Zafeiriou 2001:90). Phillips (1997:26) claims that lack of engagement leads to learners merely clicking the next screen without reading all the content on the current screen. According to Zafeiriou (2001:93), learning facilitators should make interactivity an integral part of the design rather than an afterthought. Interaction should not be used as an attention-attracting device, but should be used for educational purposes (Phillips 1997:26).

Some computer-integrated courses are designed in a traditional way that lacks interactivity (Canas 2002). In such instances, a textbook or face-to-face lecture is far better than a computer-integrated course that does not engage the learner (Phillips 1997:26). Learners understand the metaphor of a book and it is easy for them to leaf to important sections. Books contain useful visual cues that enable the learner to easily find a relevant piece of information (Phillips 1997:26). In the traditional class-room, learners and lecturers can ask one another questions and thus learners will be involved (Canas 2002).

According to Hall (1997, cited in Lautenbach 2000:26), some recent computer-inte-grated courses are too rigid. This indicates that not all courses should be delivered through the computer and that certain topics need a personal touch. Canas (2002) also warns against delivering courses online as electronic books ('dumping' text-books online) because this leads to poor interactivity and boredom. In addition, learners will not be able to deal with huge chunks of information on the screen (Zafeiriou 2001:90).

3.7.4 Computer access

Palloff and Pratt (2001:61) believe that lack of access to computers and the Internet is a serious stumbling block to successful CIE. Although organisations usually pro-vide their employees with access to computers at the workplace, most employees in Africa do not have computers at their workplaces and homes. They cannot par-ticipate in online learning activities or revise the online course in their homes (De Klerk 2002).

According to Nxasana (2002) and De Klerk (2002), current estimations indicate that the number of people who have access to the Internet in Africa is just a few million. The majority of these people are in South Africa. This makes it difficult to maximise the benefits of CIE in Africa. This could have an impact on the performance of our organisations. 'Once again, how is Africa supposed to compete fairly on this basis in a globalised world where access to the Internet and modern telecommunications is one of the basic requirements for success?' asks De Klerk (2002).

In fact, according to McCormack and Jones (1998:22), the most mentioned disad-vantage of CIE is lack of access to the Internet and computers for both learners and learning facilitators. Even if learners have computers that are connected to the Internet in their homes, Internet costs would be a hindrance (McCormack and Jones 1998:22). This could also lead to anxiety that would affect learning negatively.

Ankiewicz (1995:253) reports that in many parts of the world, especially Africa, learners hardly have access to electricity, 'let alone computers and Internet'. In some places access to technology is a time-consuming struggle that negatively affects on-line learning (Nxasana 2002). McCormack and Jones (1998:22) state that although there is a spread of computers that are linked to the Internet in developed countries, learners and learning facilitators still experience a certain level of difficulty in access-ing computers and the Internet. It is therefore important for corporations in South Africa and worldwide to consider the issue of accessibility when they offer training to their employees online. In fact, several initiatives in South Africa are being under-taken to increase the accessibility of communities to computers.

3.7.5 Copyright and privacy problems

According to Persaud (2001:109) and Murdoch (2001:34), copyright and privacy are problematic in both the physical and the virtual world. Some authors have suggested that existing laws should be applied to the digital environment (Garten 2001:28; Kern 2001:5). The problem is that countries do not have the same laws of copyright and privacy, and the virtual classroom and access to the Internet are translational (Miller 2001:59; Minkel 2002:53). In addition, some current laws do not cover the digital environment sufficiently (Perry 2001:108; Ferguson 2001b:76).

Sometimes the authenticity of the authorship of a certain work is problematic in both CIE and traditional learning environments (McCormack and Jones 1998:24). The problem of authenticity is exacerbated because the Internet has vast information and it is easy to copy (Mamudi 2001:120). In South Africa the government introduced the Electronic Communications and Transactions Ltd Act, no 25 of 2002. The ECT Act protects both online content producers and users. It also covers security-related issues, and thus these issues are not merely dictated by technicalities, but are legislated. The act reflects specific definitions of the nature of the security the organisation should provide for its website. The King II Report on Corporate Governance for South Africa also addresses these issues (King Committee on Corporate Governance 2002).

3.7.6 No benchmarking or quality assurance

According to McCormack and Jones (1998:23), the Internet and its associated phenomena (including online learning) are still in the infancy stage. The bandwidth is not sufficiently capacitated (De Klerk 2002). This has an impact on its reliability, and thus learners and learning facilitators cannot always rely on websites to obtain learning materials (McCormack and Jones 1998:23, Anon. 2001a:3).

Competition among CIE content providers (vendors) leads to learning materials that are not standardised (Anon 2001a:3). In most instances, CIE materials are not customised to national legislation. For example, in South Africa learning materials should meet the requirements of OBET, SAQA and NQF (Engelbrecht, Du Preez, Rheeder and Van Wyk 2000). Because of the relative infancy of online learning (McCormack and Jones 1998:23), it is difficult to benchmark CIE courseware with well-designed materials worldwide (Anon 2001b:21).

During the fieldwork, the researcher found documents in the corporate training environment that indicated how South African organisations handled quality-related issues regarding education and training.

CHAPTER 4

TOOLS AND STRATEGIES FOR COMPUTER-INTEGRATED LEARNING

CONTENTS

4.1 INTRODUCTION

This chapter addresses issues such as the use of CIE tools, strategies for CIE and recommendations for implementing CIE in a corporate training environment. Tools include communication tools, learner assessment tools and class management tools.

Strategies for computer education include collaboration, interaction, individualisation, problem solving, active learning and blended learning strategies. South African organisations have employed various strategies for CIE and research to transform education, training and development. This chapter will demonstrate how certain factors in the South African corporate training environment affect learners' proficiencies.

4.2 THE USE OF COMPUTER-INTEGRATED LEARNING TOOLS

4.2.1 Learning management systems: an overview

In most instances, CIE tools are part of the learning management systems (Morrison 2003). A learning management system is a huge web-based software application consisting of a set of tools that centralises and automates aspects of the learning process through the following functions (Morrison 2003:174):
- Storing and delivering self-paced online courses
- Registering learners
- Maintaining a catalogue of courses
- Downloading online learning modules
- Tracking and recording the progress of learners
- Assessing learners
- Tracking and recording assessment results
- Providing reports to management

4.2.2 Tools of learning management systems

The tools of learning management systems include communication tools, student assessment tools and class management tools (McCormack and Jones 1998:354).

a) Communication tools

Learning management systems have several communication tools to enable interaction among learning participants. Each tool uses a web-based interface. According to McCormack and Jones (1998:355), learning management system communication tools include an online class bulletin board, class e-mail and interactive chat tool.

- **The discussion tool**
  Discussions in CIE take place via a discussion tool, which is also known as a class bulletin board. The discussion tool is an electronic area for posting, displaying and receiving information. It can also be used to create an important hub of communication (Jenkins 2003:74). The bulletin board offers an asynchronous group communication system that is open to the particular learning community (Adkins 2003:28). The learning facilitator can delete messages from the bulletin board if need be. This tool is essential for collaborative learning and virtual communities.

- **The mail tool**
  The CIE mail tool is similar to e-mail and essential for learning-related private communication within the online learning course. The learning facilitator and learners are able to send, read and search for messages in the mail tool. This tool enables learners to be responsible for one another's learning as well as their own. It also allows responses, rapid feedback and the exchange of information through e-mails (Moore 2003).

- **The interactive chat tool**
  The CIE e-mail tool provides a chat device to learning participants. It is very useful for synchronous communication (Andrian 2000:110). In most instances there are more than five chat rooms, four of which are recorded and can be used for specific topics for discussion (Morphew 2000:11).

b) Learner assessment tools

The learning management system provides several tools that can be used in assessing the learners' progress (Weller 2002:116). According to McCormack and Jones (1998:355), these include self-tests, timed quizzes, learner progress and participation tracking, results management and an assignment tool.

- **Self-tests**
  Self-tests are mostly multiple-choice questions that are automatically marked by the learning management system (Weller 2002:118).

- **Timed quizzes**
  Timed quizzes lead learners through a range of questions within a certain time (McCormack and Jones 1998). Once learners have completed the questions, answers are made available to the learning facilitator, who marks and returns them via the web.

- **Learner progress and participation tracking**
  The learning management system has a tool for tracking learner participation and the number of contributions to virtual classroom activities (Gold 2003). This tool is used to ensure that learners have satisfied certain minimum involvement requirements. The progress tool also enables the learner to track his or her own progress within the course (Morrison 2003:109).

- **Results management**
  The learning management system provides an online mark book that can be used to record learner results in all kinds of online assessments. This tool has facilities for modifying and viewing learners' results, such as distribution graphs (McCormack and Jones 1998:356).

c) Class management tools

Computer-integrated class tools provide the facilitation of learner lists and results (Weller 2002:117). They manage the learner database. Tools add more columns, batch registration, select entries on various conditions and remove entries (McCormack and Jones 1998:356).

- **Learner results**
  The learner results tool is responsible for releasing learner results. It allows learners to see columns from the database. This tool can deliver results directly to learners.

- **The online marking tool**
  Quizzes are automatically marked once learners have completed them. Learners can access the 'my grade' icon to view their marks.

- **The compile tool**
  This tool allows learners to compile a set of notes from the course content and print them out.

- **The assignment tool**
  This tool contains the description, due dates and mark allocations for learning assignments (Weller 2002:125). Learners can submit their assignments by uploading them into the assignment tool.

- **The calendar tool**
  The calendar tool is shared by learners and learning facilitators. It allows private and public entries to be added. The global calendar merges all calendar entries from the various courses.

- **The learner tracking tool**
  The learning management system provides several analytical methods in graphical and tabular formats. This tool can allow the facilitator to award marks on the basis of the learner's contributions to online discussions and debates (McCormack and Jones 1998:356).

d) Use of learning management system tools by facilitators

Almost all the tools mentioned by the facilitators in this study were part of a certain learning management system. These included communication tools; discussion forums; global message broadcasting; live chat and news articles. Facilitators also used assessment tools such as 'skills assessment; test, survey and evaluation authoring; and supports certification'. Learning class management tools were course catalogue management, manager reviews and approval, resource file management, transcripts and records.

Facilitators mentioned certain tools that they used for support: 'role creation and access definition; support online; live and virtual training; curriculum path and learning programs; enterprise systems; and support for other South African languages.' The effective usage of CIE tools helps to reduce the number of learners who 'drop out because they feel isolated'. The voice tools help to 'resolve some of the social isolation problems'.

- **Learning management system tools**
  Facilitators used learning management systems to allow learners to 'log into a course, track their progress, show them what courses are available'. Some

facilitators used very sophisticated systems 'that have content management engines, assessment engines, workflow and analysis'. Facilitators and experts were mostly interested in the communication and collaboration components of the learning management systems, because they put 'your whole learning environment underneath one umbrella'. The advantages of these tools are not confined to the learner and facilitator, but operate 'between the learners themselves'. Learning content management system tools were very useful because they allowed the facilitators to manage the content, including the ability 'to allocate certain learners to certain content' or to' check where the learners are in the content'. The learning management system was also used 'to track learner usage and to report assessment results'. Other facilitators felt that the use of the online learning tools should be maximised by adding the voice element: 'People that I've been e-mailing to for the last year – you can suddenly hear them speak and it turns them into people.'

- **Integration with other applications**
  To enhance the development of human resources in the corporate environment, managers and facilitators in this study tried to integrate the learning management system tools with other applications: 'We developed an electronic performance support system to provide call centre staff with online guidance whilst dealing with customers.' According to managers and experts, there should be direct integration with the human resources 'in terms of competences and capabilities, training events, events planning, costing, so forth'. They felt that the integration of CIE with electronic performance support systems was yielding good results because 'sales consultants, with direct access to products information, could answer queries immediately and not have to keep the customer on hold or call back at a later stage'.

  The integration of CIE tools with other applications puts 'all product information at your fingertips – takes away memorising everything'. These initiatives integrate CIE with employees' work: 'The e-course could be used as reference guide for the more experienced people and an induction guide for the new people within the Call Centre.' Linking online learning tools with other systems enables managers to retain skills and knowledge within the organisation: 'There are increased product skills despite high staff turnover.' In fact, experts interviewed believed that the stand-alone systems 'don't have a future in this country'. They felt that integration of the learning management system tools would 'function as management of a whole host of things that concern learning'.

  In some industries, particularly the banking sector, there are moves to integrate online learning tools with 'knowledge management applications and e-business strategies'. Endeavours of this nature are supported by authors such as Morrison (2003:203).

e) Multimedia elements

Although some learning facilitators in this study preferred to use multimedia – graphics, tables, screen shots, illustrations and multimedia elements – 'to provide an attractiveness that assists with delegate engagement' – they found it difficult to provide this 'without either confusing the learner or simply providing these components without relating them to the learning objective'. They felt it was easier to use these elements in face-to-face situations because they could then be introduced into the session at the facilitator's discretion at the right time. In online learning the inclusion of these components in the learning material requires good instructional design; otherwise they may lead to complexity, distraction and inflexibility. They may also 'increase the risk of the content not achieving the desired end'. Graphics should not be included just for the sake of being fashionable. They should add value to the learning experience: 'A graphic should be there to create understanding and facilitate learning, not to make something look pretty.' This view is supported by authors such as Alessi and Trollip (2001:68).

Some facilitators used graphics, tables, screen shots and illustrations to attract the attention of learners: 'It's very important to have [them] there, so that people are not bored when they learn.' They felt that including these components would obviously be of benefit. Some facilitators regarded graphics as critical elements in the learning process: 'You have to have a picture on every page. You have to. What's the point of using the Internet if you're not using the multimedia?' In fact, some facilitators believed that there should be more of these multimedia components than the text: 'There is no point in putting text on the page. Diagrams, pictures, video clips and sound can convey so much more than straight text.' There was a feeling among facilitators that 'a picture speaks a thousand words'. They believed that it was content, not the multimedia elements that distracted learners: 'Some of our courses have more text. The students get tired and maybe get distracted as well. But if there's graphics, sound and animation – it keeps their attention. It's interesting to have graphics.'

f) Bandwidth constraints

The facilitators' usage of multimedia was negatively affected by bandwidth constraints: 'We don't overuse those kinds of things because we have severe bandwidth constraints.' Bandwidth seems to be a serious problem among the facilitators and managers in the banking sector: 'Our bank as you can understand, the bandwidth is there for transactions. So training is normally allocated the very last.' Another facilitator in the banking industry put it this way: 'Graphics take unnecessary space on the server – ours are used sparingly and we do not make use of multimedia and screen shots.' Some facilitators attributed the slowness of the Internet to the usage of multimedia elements: 'The Internet is too slow. Bandwidth is a huge issue.' An author who supports this issue is Morrison (2003:150). In contrast, however, some experts did not believe that bandwidth was such a serious issue: 'I don't think the bandwidth is necessarily such a problem. I just think that learning content should be developed according to the bandwidth.' Many effective interactive learning programs are developed 'with very low connectivity speed. I don't think technology is as much of a barrier as people make it out to be'.

g) Learners and multimedia elements

Learners commented that graphics made the lesson interesting: 'I think I prefer graphics.' 'If there's only writing you got to read; it becomes boring.' In fact, they found that graphics and illustrations 'add value because they make you fully understand what you are learning'. Some learners felt that the multimedia elements should not be exaggerated because 'there needs to be a balance between something like – between bells and whistles with drag and drop templates'. Learners noted that there were online tools but they were not being fully utilised. Learning facilitators who were part of focus group interviews with learners confirmed this: 'There is that feedback option too' and 'we tend not to do it'. Learners also found that some tools were not used for the purpose for which they were bought: 'We haven't begun to optimise the possibilities of the system we have; we have a downfall in the usability of it.' The underutilisation of tools led learners to ask the following questions: 'How do you get to handle immediate questions? How do you tend to handle long-term questions?'

4.3 RECOMMENDATIONS

- All organisations should try to integrate their learning management tools with other applications within the organisation.
- Corporate South Africa should liaise with government and government enterprises such as Telkom in dealing with the issue of bandwidth constraints. This would enable facilitators to maximise the use and benefits of multimedia elements without being hamstrung by bandwidth limitations. In fact, cooperation between these stakeholders is long overdue. CIE practitioners in other countries regard the bandwidth issue as history, perhaps because of cooperation between stakeholders in those countries.
- A lot of support should be offered to facilitators so that they can become more familiar with the learning management systems. Lack of familiarity perpetuates the underutilisation of CIE tools.

4.4 STRATEGIES FOR ONLINE LEARNING

4.4.1 Collaborative learning

According to Gokhale (1995) and Panitz (1996), CIE encourages cooperative and collaborative learning. Learning facilitators can also create virtual classrooms and communities, supplemented by Internet services such as e-mail.

According to Gokhale (1995), the term 'virtual community' refers to a group of people who engage in online collaborative learning. The successful performance of one learner has a positive impact on the whole virtual community. A virtual community is therefore a platform where a group of participants share common learning practices, are interdependent, make decisions together and identify themselves as a learning community (Crook 1999:70). This kind of learning allows learners at various performance levels to work together towards a common learning goal. Botha (2000:24) reports that learners are capable of performing at higher intellectual levels in collaborative situations than when required to work individually. Group diversity in terms of knowledge and experience contributes positively to the learning process. Panitz (1996) recommends that a virtual community should be characterised by content-free ways of organising social learning and integration in the group. This kind of learning should transcend the classroom. Nevertheless, in the ideal virtual community, the facilitators, by virtue of their authority as professional practitioners, make the primary decisions concerning the learning product and assessment, with the views of other members of the group not being subordinate to those of the trainer (Tucker 1997:31).

According to Bonk and Cunningham (1998:40), virtual communities are open systems and international platforms in that they allow the participation of learning facilitators, managers and learners outside the system by means of external links. Books, however, are closed systems, predetermined by the course designer. The openness of the World Wide Web thus further supports CIE (Hannafin and Peck 1988). This openness allows members of the virtual community to participate in online learning activities while on holiday or on a work assignment anywhere in the world. Learning facilitators can create online courses that are globally accessible (Hannafin and Peck 1988). As increasing numbers of courses are published on the World Wide Web, collaborative learning is enhanced. In this sense, CIE can be described as non-discriminatory, because virtual communities are accessible to all learners and trainers, regardless of their situation, age, ethnicity, gender, language or physical limitations. Courses may be set up in a way that makes CIE more anonymous and therefore less inhibitive. The virtual community may be experienced as being more secure, and learners may therefore be confident and participate in its activities without being shy (Tucker 1997:30).

Only a few of the organisations that participated in this study enabled their learners to take part in collaborative learning, although the learning management system and/or virtual classrooms in all these organisations catered for collaborative learning: 'Truly speaking we don't really use collaborative learning.' Facilities included online chat rooms and discussion forums: 'In this company we do make use of collaborative learning during the asynchronous sessions.' 'We are not sure if employees will want to participate in synchronous sessions.' Some learning experts believed that research should be conducted 'to investigate the efficacy of collaborative learning in the corporate training environment'. However, research conducted in universities and schools shows 'that collaborative learning can add a lot of value in teaching and learning activities'.

Learners found that CIE enhanced collaboration. In most instances, collaborative learning took place during the synchronous sessions 'where they actually had an interaction between themselves'. Participants really liked these kinds of sessions: 'I thought that went down very, very good.' The advantage of collaboration during the synchronous sessions is that 'you don't only have one person speaking and everybody else listening'. Collaboration brings the human touch to online learning. In its absence, online learning is not user-friendly 'and I think that is a mistake number one'. CIE has to 'be more people-focused, because it's not'. Learners in this study stated that they wanted more collaborative learning because 'sometimes it does lose that human touch' and they 'would like the human touch'. Learners also wanted collaboration in asynchronous sessions 'because you have questions and there's no trainer there to ask, so there's a bit of a shortfall when it comes to online learning'. A collaboration strategy was a solution to this: 'Within this system we can communicate with the different users who do the same courses.'

4.4.2 Interaction

Effective CIE should increase interaction between the learner and the tutorial (Hannafin and Peck 1988). Interaction is the active exchange of information between the learning facilitator and the learner. The learning program should present the learning content, and the learner should respond. The learning program will then decide its course of action, guided by the learner's actions, and the process is continual (Hofmann 2003a).

In a face-to-face learning environment, the learner may lose concentration or deliberately tune out if she/he is bored (Hofmann 2003b). However, in CIE, progress through the learning program should be tied directly to the learner's actions. If the learner daydreams, the learning program waits patiently (Hannafin and Peck 1988). Then again, if the learner has not learned sufficiently, she/he should be given additional instructions. CIE should provide appropriate interaction to encourage and maintain learner participation (Hofmann 2003).

With CIE, the learner can go through the lesson at the pace that suits him/her (Tiffin and Rajasingham 1995:143). CIE gives the learner power and a feeling of total control (Wolf 2002:13). As a consequence, learners learn quickly because they are at liberty to choose where to learn. Because CIE gives the learner freedom of time, space and pace, this makes learning more productive and less stressful (Tucker 1997:30). CIE builds up the learners' confidence. Learners can also become explorative by going through lessons that use artificial intelligence. During exploratory learning, the learner is not shy to make mistakes because he/she is learning in private and will not look stupid by answering wrongly (Tucker 1997:30).

Facilitators used online learning to increase interaction between the learners and the learning program. According to some facilitators, interaction is very important in the learning process and thus 'interactivity is the buzzword'. To be successful, interaction should be preceded by a great deal of work: 'That means you have to have trained instructional designers who help. You have to have people that create proper web pages.' Facilitators used learning management systems that are interaction-friendly. In the absence of such systems, 'you've got to have Java Script programmers and graphic people'. In fact, to initiate a fully interactive 'online course involves something like seven kinds of specialists'. Learners and facilitators in this study were committed to maximising the benefits of interaction by setting aside time for their studies. Yet 'this is one of the major problems' that is being addressed by what they call 'learner contract'. Facilitators should also make certain commitments to their online learners: 'This is what I'll do for you. I will always respond to your e-mails within 24 hours. I will always give you help, but you must also put aside one hour a day to work through this course.' Interaction as a strategy needs the active involvement of facilitators, learners, designers and programmers. Morrison (2003) agrees with this argument.

Learners found that CIE increased interaction between the learner and facilitator/ learning program: 'With e-learning you can get what you want to know immediately. If you need to know more you can ask for assistance and get immediate feedback.' Learners found that online learning provides feedback constantly and thus 'most of the time – I'll say 99% – it is accurate'. Learners have realised that they may not always be able to ask questions in traditional learning, 'and even if you can, you need to have an answer quite quickly and that's a problem'. However, with CIE 'you can ask a question and get an answer straight away' and thus momentum in the process is not lost.

4.4.3 Individualisation

Although interaction is regarded as a primary contributor to the effective CIE program, individualisation should mostly be used for its efficiency (Hannafin and Peck 1988:8). The one-to-one nature of CIE programs should be used to monitor learner understanding frequently and to respond according to the needs of each learner (Porter 1997:27).

According to Hannafin and Peck (1988:11), a possible strategy is to start with a pre-test. The pre-test may be used to determine whether the learner has the pre-knowledge that will enable him/her to be successful in the learning program and to confine the lesson to specific modules that he/she requires. Once the lesson on the required topics has started, the learner's actions are used to ascertain when mastery has occurred, so that the instruction may proceed to the next topic (Weller 2002:13). Remediation and additional instruction should be provided if they are necessary (Alessi and Trollip 1991:71).

Hannafin and Peck (1988:9) encourage computer-integrated instructional designers to accommodate a certain element of personalisation. For example, personalised reference to learners in a lesson will make the instruction more interesting, more relevant and more effective. This can be achieved by using learner's name in assessment and feedback.

According to Alessi and Trollip (1991:71), in the traditional learning context, the instructor will not be able to provide feedback at the same speed as the computer-integrated program. Educators should therefore employ CIE to provide assessment and feedback after a very short period (Weller 2002:72). The capacity to provide immediate feedback is a key factor in CIE strategies (McCormack and Jones 1998:236).

Tucker (1997:24) claims that because employees receive a percentage of what they hear, they move out of the classroom with a percentage of what the instructor has presented. Traditional training seldom conducts post-training assessment. This problem is exacerbated because the trainer does not know how much the employee has learned, and the learner does not know how his/her knowledge and skills compare with the expected level (Zafeiriou, Nunes and Ford 2001:87). Hargis (2001:477) reasons that self-reflection is the most important characteristic in learning. This includes perception of self-efficiency and capacity to organise and carry out actions required in learning and working environments.

CIE interactivity allows the learner to grasp the required knowledge and skills (Tucker 1997:24). Learners are also assessed to ensure that the required level of knowledge/ skills has been met and, if not, the learner is directed back to the sections where knowledge and skills do not satisfy the required standard (Lindroth 2002:23).

Facilitators employed self-paced learning (individualisation) because 'real learning always takes place within the person, within his or her brain and emotion'. Facilitators also used individualisation because an average employee 'wants a comfortable way and a convenient way to study', and that is what online learning provides learners. CIE is characterised by 'a lot of interactivity built in, a lot of multimedia to keep the lesson appropriate to your learner who now essentially may be studying on his or her own' to keep him or her motivated. Facilitators in a mining organisation noted that 'the delivery of individualised and dynamic online learning material has proved successful as part of the diamond mining giant's leadership development programme'. According to the facilitators, learners have seen the benefits of self-paced learning: 'Their attitude is very good.' They really want to learn and 'they are self-directed learners'. 'The employees could work at the right time at the right place' and this enhances individualisation.

Learners found that online learning enabled them to study on their own: 'I see it as important, very much self-directed learning.' Learners also enjoyed the convenience of CIE: 'It's very convenient to use online learning because I can set my own pace.' This convenience allows an individual learner to learn wherever he/she is as long as he/she has access to a computer and the Internet: 'You don't have to travel to other institutions.' Learners can study in their own time: 'You can take your own time and if you don't want to do it now, you don't have to do it now.' Learners were excited about the benefits of individualisation because 'it's me in my own little world'. Individualisation allows learners to make mistakes without any fear of embarrassment: 'You study on your own time rather than studying with people. You feel embarrassed when asking some questions. You are afraid to answer the questions.' However, 'if you are alone, you can do it in your time. That to me is really very good.'

4.4.4 Problem solving

James Kullik (in Molnar 1997:66) states that CIE could raise learners' functioning by 10 to 20 per cent and moderate the time set aside to attain a particular learning goal. Tucker (1997:25) also remarks on the improvement of the general performance of learners. As learning tools, high-powered computers have revolutionised the arrangement and manoeuvring of content. CIE has extended the wits and insight of learners and their learning facilitators (Molnar 1997:67). While people find it hard to deal with enormous quantities of data and information, computers can do so effortlessly, allowing learners to focus on learning rather than struggling to deal with information.

Littleton and Light (1998:2) contend that collaborative learning can augment cognitive development and learning performance. Where many people engage in a learning task, even where group partners share the same level of cognition, each is bound to keep in mind different sections of relevant information. These individuals will bring together their different talents to confront the problem. At the end of the task, all members of the group will have gained, including the learner whose participation has been low.

The scaffolding provided within the web system requires learners to extend their thinking in order to unravel the problem. There are still many instances where trainers try to develop learners' critical thinking in traditional and formal educational settings, but the online setting provides new opportunities through communication, information resource access and joint function.

In most instances, interactive CIE precludes rote learning because it requires critical thinking (Crane 2000:43). Critical thinking in turn requires learners to scrutinise, organise and synthesise information. Interactive CIE will therefore advise a learner to move ahead of what he/she is reading.

If learners are to cope with CIE, they should be able to think critically and creatively. Crane (2000:43) refers to research conducted by the National Assessment of Educational Progress which showed that most learners in the traditional learning environment do not show any significant development of higher-level thinking skills and cannot therefore apply problem-solving strategies, possibly because many organisations expose learners to the conservative and formal education situation only. Once in the working environment, these learners are consequently criticised by employers for their inability to apply critical and creative thinking skills (Crane 2000:44).

An important advantage of incorporating e-learning is that this teaches learners to use computers as tools for finding information. It is better to teach learners how to discover, synthesise and organise information than to force them to learn facts and concepts by heart. CIE interactivity and its resources assist learners to develop thinking skills that prepare them for many complex situations they will encounter in their future. Mendrinos (1987, in Crane 2000:44) supports this analysis, stating that online learning encourages learners to search and retrieve information; through these activities, their deductive reasoning and critical thinking are developed.

4.4.5   Active learning and constructivism

Constructivism is a recent philosophy of learning that has implications for CIE (Bonk and Cunningham 1998:32). A major tenet of this philosophy is the active nature of learning, which grows out of the learners' experiences (Alexander and Boud 2001:7). This philosophy is compounded, because trainers cannot transmit knowledge and meaning, but learners should construct it themselves (Hargis 2001:477). Knowledge is not something people possess in their heads, but is constructed in their minds (Ludlow et al. 2002:38).

Phillips (1997:21) and Van der Vyver (2000:39) suggest that rather than identifying the set of skills to be transferred into the learners' heads, attention should shift to establishing learning environments that are conducive to learners constructing their subject settings. This could be achieved by integrating CIE in a working environment. Hargis (2001:480) claims that CIE has more advantages in terms of enhancing active learning and constructivism than traditional learning. CIE enables learners to learn constructively by interpreting, discussing and applying the learning content (Chambers 2002:9). CIE enables the trainer to create a constructive learning milieu for his/her learners so that boredom may be conquered (Bonk and Cunningham 1998:33). CIE could give the learner a rich learning environment filled with authentic problems (Phillips 1997:21).

According to Johnson (1997:170), as learners interact with others 'in a contextually rich learning environment, they pick up relevant jargon, imitate behaviour, and gradually start to act in accordance with norms of cultural setting'. CIE enables learners to construct knowledge by bringing them together through electronic chat rooms and discussion boards. In a group with a high morale and a positive class spirit, even the weaker learners could be inspired to achieve more.

During constructivist learning, learners can also assume the role of explorer/inventor (Wheatly 1991; Quilling, Erwin and Petkova 1999:6). The learning facilitator in such a situation should be viewed as a valuable resource who facilitates learning rather than acting as an authority who dominates the learning activities (Carnevale 2001:43). Through experimenting, the learner can actively learn much about any subject matter (Johnson 1997:171).

Some of the CIE offerings were directly related to the employees' work activities (co-operative education). This allowed the facilitators to enhance active learning: 'We've got some induction program on e-learning and then we've got basic safety programs at our mines.'

Active learning adds value to the employees' work activities, and that is why in one organisation, before the employees did a safety and health training course online, 'there were lots of fatal accidents a month. But now it is reduced to three fatal accidents.'

In a certain bank, before call centre employees did an online course that focused on their work, 'too few customer enquiries were converted to sales; there were incomplete customer requirements to sales staff; there was high staff turnover and limited knowledge of the bank products', but after the online training 'the level of call centre customer service improved; more customer enquiries were converted into sales and product knowledge increased'.

Facilitators used online learning to enhance the skills of employees who perform technical jobs: 'Online learning programs are also being offered to blue collar workers, such as factory foremen and artisans.'

Learners found that CIE enhanced activity-based practice and cooperative education (work-based education) because some courses were directly linked to their job descriptions: 'I find that there are a lot of courses that you can choose from. The ones that I've accessed are the ones that help me with my work.' Learners were very impressed with the linkage of online courses and their work activities: 'It is interesting for me as well', 'I do courses that are more applicable to my work.' Perhaps this is why some employees prefer to work and learn simultaneously: 'I can actually switch between the software program on my computer and go and try it quickly and go back to the course – carry on with the course.'

Some online courses were directly linked to the organisation's business requirements: 'I learned that it's very important – e-learning to employees – because it has to do with the running of business. Especially the courses that they put on the system are very, very important.'

In fact, learners found that CIE courses made their job easier because they were now more skilled: 'Now my work becomes easier – because some of the things that are dealt with are about helping the customer. Now I've developed.'

a) Strategies

The major classifications of CIE strategies include tutorials, drill and practice, educational games and modelling/simulation (Hannafin and Peck 1988:139).

- **Tutorials**
  Hannafin and Peck (1988:139; Dagada and Jakovjevic 2004:195) claim that in tutorials the information is presented, verified and reinforced through interaction with the CIE program. Computer-integrated tutorials are usually employed to present new information and certain skills and concepts to learners (Porter 1997:87; Hill 2006). Alessi and Trollip (1991:17) and Young and McSporran (2004:350) emphasise that learners should be assessed during the tutorial to ensure comprehension. Hannafin and Peck (1988:139) declare that an efficient tutorial should include orientation, guidance throughout the lesson, feedback, remediation and strategies for making the lesson meaningful to the learners.

- **Drill and practice**
  Hannafin and Peck (1988:139), and Woit and Mason (2000:368) argue that much of what learners should be taught needs practice to promote proficiency and comprehension. Drill and practice give an educator an opportunity to determine what learners are able to do, target skills, give feedback regarding proficiency and remediate those skills that learners are not performing well (Alessi and Trollip 1991:91; Sein and Simonsen 2006:138; Savidis and Stephanidis 2005:43). According to Siegel and DiBello (quoted by Hannafin and Peck 1988), drill and practice, feedback and remediation are partners in the teaching and learning cycle. Hannafin and Peck (1988:144), and Adams (2006:25) warn against misapplication of drill and practice. Misapplication could be eliminated by providing immediate feedback and focusing on learn-

ers individually. Immediate feedback enables learners to make corrections and/or reinforce correct responses. It also adds value to positive instruction (Weller 2002:72).

- **Educational games**
  Games have been utilised for many years as a learning motivational strategy (Hannafin and Peck 1988:154; Lee, Luchini, Michael, Norris, and Soloway 2004:1376). Online learning games are also used to motivate learners (Alessi and Trollip 1991:162). They are generally used to reinforce the information, skills or concepts that have already been taught. They also tend to help learners develop and reinforce and refine certain aspects of learning (McCormack and Jones 1998:236; Halverson, Shaffer, Squire and Steinkuehler 2006:1049). Hannafin and Peck (1988:154,) and Denis and Jouvelot (2005:463) stress that the designer should refrain from using games in online learning like non-educational computer games. The primary goal of games in CIE is to add and retain educational value and apply acquired skills (Hannafin and Peck 1988:154; Fisch 2005:57; Zheng and Young 2006:874). These games may be characterised by self-directed competition, graphics, sound and motion (Ke 2006: 315; Alessi and Trollip 1991:162).

- **Modelling/simulation**
  Computer-integrated simulations estimate, imitate or copy the characteristics of some task, setting or context (Hannafin and Peck 1988:149; Brown and Lahoud 2005:65). Chernobilsky, Nagarajan and Hmelo-Silver (2005:53) indicate that simulations are employed when the costs of optional teaching methods are very expensive, when it is not possible to learn the concepts of interest in real time, or when competence needs to be demonstrated in a controlled risk-free setting. Online learning simulations represent an approximation of the real context (Kazmer and Haythornthwaite 2005:8; Galvao, Martins and Gomes 2000:1693).

Hannafin and Peck (1988:149) note that 'simulations are used to study the effects of the experiments that cannot be observed by the naked eye, due perhaps to the microscopic nature of the experiment or to the extreme speed or slowness of the elements to be observed'. It would be difficult to observe the real splitting of an atom in a nuclear reactor, and thus the only possible manner to illustrate this process is to simulate the splitting through a computer-integrated program. In this instance, the process could be slowed down enough to allow careful study (Musslewhite 2003; Kazmer 2005:8). The same method could be used for processes that need long durations of time and that cannot be observed in real time for classroom teaching and learning. It is, however, very important for the designers to create simulations in a way that provides scenarios, completion simulations and clear options for learner participation and guidance (Alessi and Trollip 1991:119; Kazmer 2005:26; Jain and McLean 2003:1069).

Facilitators used various modes and designs of CIE such as tutorials, simulations and drill. These strategies enable the facilitator to deliver the message and the learner to comprehend. Learners become involved in the learning process, and thus interaction between them and the computer-integrated program is enhanced. Facilitators stated that they used drill when they wanted learners to be acquainted with the 'speed and accuracy' of doing certain tasks. The repetition enables learners to be more accurate and more conversant with the speed. Once learners 'fully comprehend the speed and accuracy of certain tasks, they can proceed to other tasks or learning programs'. Simulation was used when 'the real demonstration of certain tasks will be dangerous, dull and difficult'.

A CIE expert gave an example of a South African air transport provider which was using online learning simulations and tutorials to train 'about 10 000 of its employees this way so that they will be acquainted with conversion to the airline's new Airbus fleet'. The carrier was in the process of replacing its existing Boeing airliner fleet with new state-of-the-art airbuses. Tutorials were used to 'provide certain facts about the Airbus', while simulations were used to 'provide demonstration of certain things about the Airbus'. The tutorials and simulations were not confined to the airbus conversion, but were used 'to demonstrate medical, safety and communication skills training to flight deck and cabin crew'.

Learners who participated in this study found that CIE employed various modes and designs such as tutorials, games and simulations. These modes are used together with multimedia elements: 'There are pictures – even in new courses we've got flash animations.' These elements are used to demonstrate certain aspects: 'The A+ courses – they show you how to disassemble a computer screen.' This allows learners to 'actually see what's happening, which is interesting'. According to the learners, there were 'clear examples of things' the facilitator could use to retain their learners' attention.

4.4.6   Blended learning

CIE has rapidly found its way into the corporate training environment. However, a trend that is emerging is that online learning is being integrated in other training methods (Van der Westhuizen and Krige 2002). These include the traditional face-to-face models (Troha 2002).

Blended learning is a training solution that mixes several delivery methods, such as face-to-face training, CIE and self-paced learning (Valiathan 2002). A blend can also refer to a delivery mode (Valiathan 2002; Hofmann 2001). Smith (2001) outlines the following examples of blended learning:
- Traditional workshops or seminars in conjunction with a teleconference feature
- Traditional courses with a continuing e-mail connection or ongoing dialogue with the participants
- A traditional seminar with live broadcasts to more than one site

Burrows (2002:15) and Allerton (2002:6) remark that blended learning is touted as the latest panacea for corporate training. However, Saunders and Werner (2002) and

Barnum and Paarmann (2002:57) note that the employment of multiple strategies, methods and systems make up for the weaknesses in any training methods or delivery mode.

Almost all facilitators used CIE with other learning delivery and pedagogical methods such as traditional face-to-face learning, workshops, paper-based learning materials and instructor-led sessions. In fact, some facilitators and experts did not believe that 'the excellent teacher that stands in front of a small group of people in front of a blackboard with a piece of chalk in his hand', taking learners through an 'intricate, say, accounting exercise' would ever be replaced by CIE, and thus it was important to mix CIE with other traditional methods.

A facilitator who wanted to teach his/her learners a high level of, for example, a certain therapeutic approach 'would give them a lot of up-front reading to do. That, they would be able to do online. They would be able to access my notes on that, they'd be able to access the World Wide Net, to access my notes on articles that I'd scanned specifically for this group' so that they could prepare themselves on a theoretical level, 'but then, in the end, I would like to have them in small groups, eight people, working in a training room, in a workshop where there'd be role play, where I show them a video of a therapeutic session being conducted, where we can discuss it'.

The nature of the blended strategy employed in a certain lesson depends on the content. Facilitators, for example, gave learners paper-based handouts or printed out the documentation: 'When it's really intricate, when it's really difficult, when I really want to apply myself, when it's complicated, I still have to have a printed paper in front of me and be able to work through it with a pen or pencil in my hand.'

Learners realised that the presence of CIE in their organisations lent itself to blended learning strategies: 'Proper target analysis has to be done to see how you can cater for your blended learning styles.' Learners tend to grow tired of a single delivery mode: 'I don't know if you've heard the saying "Death by PowerPoint". You stay in a classroom situation and the trainer stands there and just flicks through his Power-Point presentation – there's no interaction.' 'Death by PowerPoint' is not confined to face-to-face training, because CIE can also become a victim of this: 'Almost "Death by Sentra" because the facilitators don't know or haven't got the skills yet.' On these grounds, organisations are employing blended learning strategies.

4.4.7 Research as a strategy

South African organisations, especially in the banking sector, employed research as a strategy to improve the use of CIE. Documents show that almost all organisations that use CIE conduct some sort of action and applied research to improve the integration of CIE. The Banking Sector Education and Training Authority (Bankseta) is spending a lot of money on research activities. In 2003, the Bankseta conducted an online learning benchmarking exercise by sending a fact-finding delegation to the US and Canada. The mission had representatives from Absa, Standard Bank, Nedbank and First National Bank, the Bankseta and the South African Reserve Bank.

South African banks were able to learn from and benchmark themselves against their counterparts internationally regarding CIE in the corporate training environment (Bankseta 2003a). The lessons learned were shared with other banks and organisations that are not in the financial sector. The Bankseta also awarded an e-learning feasibility study to the University of the Western Cape (UWC) and Safmarine Computer Services in 2003. This research project, which was completed in October 2003, focused on the possibility of implementing a sector-wide e-learning initiative within the banking industry (Bankseta 2003b). In May 2003, the Bankseta hosted an international conference at which e-learning was one of the themes.

4.5 RECOMMENDATIONS

- Only a few organisations gave their learners opportunities to learn collaboratively. Facilitators should create these opportunities. This will allow learners to share information about computer-integrated courses and other work-related issues.
- Facilitators should employ learning strategies by maximising the use of the learning management systems.
- An active and cooperative learning strategy should be emphasised by linking more generic courses with organisational and employees' needs; otherwise senior management and learners will not realise the importance of computer-integrated courses.
- A simulation strategy should be optimised, since this strategy can estimate, imitate and copy the characteristics of the real context.

4.6 FACTORS AFFECTING LEARNER PROFICIENCY

Several factors affect learners' proficiencies in CIE.

4.6.1 Managers' and facilitators' perceptions

a) Learners' attitudes

In most instances, managers and facilitators perceived learners as being positive about participating in CIE. A positive attitude affected their proficiency in a constructive way: 'My perception is that people are generally positive about going on training. They see it as a benefit and improvement.'

They claimed that the attitude of learners was enhanced by willingness to learn: 'The learners want to learn.'

They found it surprising that the attitude of learners was not affected by the reluctance of line and middle management: 'Your middle management doesn't buy in, but your learners do. There is a paradox between the two.'

Other facilitators indicated that the attitude of these managers had a direct impact on learner proficiency. However, they said that most learners wanted to learn against all the odds because 'they are self-directed learners; they want to participate'. Authors such as Hannafin and Peck (1998) and Alessi and Trollip (1991:70) support this view of the impact of self-directed learning on learner proficiency.

b) The attitude of line and business unit management

In some instances line and business unit managers did give learners sufficient time for CIE activities: However, 'about 80% of them [business unit managers] said that they won't have the time to do e-learning'. Some of these managers did not believe in e-learning, and thus in certain organisations, during the pre-activity survey for CIE, these managers said: 'I don't believe e-learning would work.'

In other organisations, by contrast, managers were very positive about the contribution of CIE to human resources development and said that it had been identified 'as our first support'. These managers actively participate in the co-ordination of CIE in their units, divisions or branches: 'They do the workplace assessments for us. They ensure that the delegates, employees complete the required content.' This view of management support is documented by authors such as Tetiwat and Igbaria (2000:17) and Boxer and Johnson (2002:37). In an organisation with branches, trust between the facilitators and branch managers is essential because they may allow learners to have access to answers during the assessment sessions: 'The other thing that you need to work around is the trust between yourselves and the line managers. They basically check assessments you use.' This could be attributed to the fact that 'there's a human element and a technology element'.

c) Mindset as a factor

Most of the managers and facilitators reported that the mindset of management and learners has an impact on CIE. In fact, mindset is crucial in learner proficiency: 'In the past we had a huge change manage problem.' Learners did not really regard computers as learning tools and were reluctant to use them: 'Up to about three years ago there was reluctance to use the computer, but we find that they see the benefit of the training – and that they can be trained much quicker than they were in the past.' Learner proficiency in CIE always has to do with the mindset towards learning and not necessarily towards online learning. On the other hand, 'learners don't understand the value of learning itself and therefore do not necessarily want to learn'.

Managers said that they were involved in introducing a culture of lifelong learning as a programme that was embedded in their organisation. Organisations were also battling with 'the transition from a social point of view'. They said that learners were used to going to face-to-face training courses where they socialised with other employees. However, with computer-integrated learning they 'sit in front of this PC and whom do they socialise with?' Authors such as Kemery (2000) agree that socialisation is an aspect to be considered in CIE.

d) Accessibility to computers

Accessibility to computers, according to the managers and facilitators, was a factor that influenced learner proficiency in CIE in the corporate environment: 'Sometimes I think students don't have computers. Then it's a little bad for them. They can't do the courses or they have to ask the managers/supervisor to use the computer in the office.' However, when learners had sufficient access to computers, facilitators thought that 'the students are doing the courses very well'. In the banking sector, the insurance industry and the IT sector, for example, most people have access to a computer.

In other organisations, accessibility depends on the division: 'People that are working in the production line or on the shop floor do not have access to computers.' Those that do, might not have regular access, and thus learner proficiency is affected. Many large corporations have learning centres, and this contributes positively to learner proficiency. Other organisations have learning institutes, campuses, virtual learning centres and computer centres.

e) The profile of the learners

In some organisations, the target 'would be the younger members of our organisation', and thus their proficiency was very good because they were 'so very young and dynamic'. They 'enjoy computers, they know computers, they are comfortable with computers'. Unlike older partners or managers, they 'don't have those [technological] constraints to deal with'. Even learning experts perceived age as a serious factor in CIE: 'My perception is that people are generally positive about going on training. Particularly at junior level.' This assertion is supported by authors such as Chau and Hu (2002) and Friendlein (2001). Sometimes facilitators redesign programs according to the age of the target population: 'There's younger learners that get more funky stuff and there's older learners that get more stuff based on computer-based training and get print-outs, together with a whole blend of solutions.'

4.6.2 Factors related to learner proficiency

a) The facilitators' competence

Learners in this study realised that their proficiency and shortcomings were influenced by the facilitators' competence in handling online courses: 'But a lot depends on the facilitators as well to engage the audience.' Learners noted that 'just because you're a good facilitator or trainer in the classroom does not necessarily mean that you're going to be a good facilitator in a virtual classroom environment'. Learners were frustrated by lack of competence of facilitators: 'Because the facilitators don't know or haven't got the skills yet, and it's very simple if you know how to and if they give you tips to get your learners engaged; but they don't at this point.' Learners stated that they had attended many sessions and that 'the facilitator is not very good' and that is why, as a learner, 'you actually start looking at your e-mails'. Learners believed that their proficiency would improve if 'the facilitator is good'. Learners found that facilitators could not 'optimise the possibilities' of the learning management system, and thus feedback tools and other online facilities were not used.

b) Pre-training

Learners were given pre-training that acquainted them with computer literacy and the learning management system: 'As a learner you go through participant training. So if you're going to attend a Sentra session or a virtual classroom session, you need to go through that training to get used to the actual environment and system.' Learners thought they were not really receiving sufficient induction before they engaged in online learning activities: 'When it comes to Topclass, our LMS where all content is stored – learners don't necessarily go through an induction there, which I also think is a bit lacking because I don't know what to expect.' Some learners, however, appreciated that they were computer-literate before they participated in CIE. Other learners, on the other hand, indicated that online learning enabled them to become computer-literate: 'What I've learnt again from this software is that it is based on computer literacy.'

c) Pre-training of facilitators

'I think trainer education – "training the trainer course" – will be more beneficial.' Learners believed that if facilitators received sufficient pre-training, it would have a good impact on their (the learners') proficiency: 'If you're a trainer you need to go on leader training.' Learners stated that they would like trainers' facilitation skills to be improved: 'Participant and leader training where they try and take you to another level in terms of your facilitation skills and also the interface changes.'

d) Recognition of pre-knowledge

It seems that pre-knowledge, or rather audience analysis, is not given sufficient attention: 'If you're a learner going to an advanced course and you might have known something. A lot of times the course forces you to go through certain aspects, and sometimes they can be tedious – if you know already.' This affects learner proficiency negatively because 'you need to take people that actually do know more and we tend to actually take the people that don't know much, which I think is where we fall short'.

e) Accessibility to computers

Access to computers affects learner proficiency positively or negatively. In most instances in this study learners had access to computers only at work: 'We do have access at work. I am using the one at work in my spare time.' The lack of computers at home may have negative implications for learner proficiency in computer-integrated courses: 'I don't have one at home.' If the learner cannot use the computer at work owing to workload, then he/she will not be able to participate in CIE.

f) Time spent on computer-integrated learning activities

The quantity and quality of time spent on CIE affects learner proficiency. Most learners had to use their spare time for computer-integrated courses: 'During the lunch breaks you can study online. You get time or you make time. You've got to study. It's only for the development of yourself.' Learners should therefore take some responsibility for their learning and development. Sometimes employees engaged in CIE activities when they were not busy. Learners noted that some of their colleagues wasted time by doing unproductive things on their office computers when they were free: 'There are lot of people that rather play games and do other things' instead of developing themselves. Learners indicated that managers were also obliged to give employees some time for learning. Line managers were 'learning promoters' for their subordinates: 'Your promoter or manager should actually give you time to do the courses, it's part of development.'

4.7 RECOMMENDATIONS

- CIE practitioners should focus on the attitudes and mindsets of management and learners since these have an impact on the integration of CIE. This can be achieved by making sure that the outcomes of CIE are visible and add value to human resources development. CIE practitioners should employ strategies to deal with resistance.
- Organisations should ensure that learners have access to the Internet and computers. Computer centres should be established in environments that do not allow each employee to have his/her own computer at his/her desk.
- Learners' pre-knowledge should be recognised before they engage in a particular computer-integrated course. Facilitators should conduct analyses of learner demographics, psychographics, attitude, experience with CIE, prior knowledge and experience.
- Human resources development managers should introduce some incentives that make facilitators and learners more committed to CIE activities.
- Human resources development managers should introduce programmes and workshops that make facilitators competent. Learner proficiency will be more positive if their facilitators are competent.
- Facilitators should give learners pre-training courses so that they can participate actively in CIE. These should include pre-training in computer literacy and learning management systems.
- Line managers should provide their staff with sufficient time to do computer-integrated courses.

CHAPTER 5

COMPUTER-INTEGRATED LEARNING AND QUALITY ASSURANCE ISSUES

CONTENTS

5.1 INTRODUCTION

To ensure that computer-integrated learning materials comply with the principles and requirements of OBE, the NQF and SAQA, the designers should adhere to good learning materials quality standards. In the South African corporate training environment, this was achieved by applying sound instructional design; implementing quality assurance strategies; considering factors that are related to learners' perceptions of quality assurance; and applying good interface design. Besides these issues, this chapter will deal with tools that can be used for quality purposes. These include a computer-integrated evaluation instrument course; a checklist for well-designed online learning materials; and criteria for evaluating quality courseware. Although these tools are mostly used in higher education, they are applicable to the corporate environment.

The organisations employed several steps for instructional design. These included analysis, design, development, implementation and evaluation. This chapter will demonstrate how sound instructional design can enhance the quality of computer-integrated learning materials. It will also show how managers and facilitators used quality strategies deduced from the corporate training environment.

This chapter will show how user interface design may be applied to enhance the quality of computer-integrated learning materials. Issues that will be addressed include the introduction of a program, learner control, presentation of information, providing help and ending a program.

5.2 INSTRUCTIONAL DESIGN

5.2.1 Analysis

In the analysis phase, data is collected about a number of aspects. These include evaluating business and instructional goals, learners and technology. Organisations used some of the tools that Kruse and Keil (2000:63) suggest for gathering information:

- A *survey or questionnaire* can be used to pose specific questions to the employees. The data is then reviewed and summarised.
- *Direct observation* enables the instructional designer to personally observe employees when they are doing certain tasks in their workplace.
- *Interviews* enable the instructional designer to be in touch with experts and to do a random sample of learners through one-on-one interviews.
- *Focus groups* allow the instructional designer to pose questions to a group of subject-matter experts and learners. Data is obtained from direct answers and conversations among the focus group participants.

a) Business and instructional goals

The first step in business analysis is to determine the desired performance goals. Business objectives should be analysed by selecting the appropriate intervention. If a training intervention is suitable, the goal should then be modified into a learning outcome.

b) Assessing learners

After the desired training has been identified, the targeted learners are analysed. Issues to be explored include those suggested by Kruse and Keil (2000:65):

- *Demographics*: What are the characteristics of the targeted learners? Is there uniformity in terms of gender, age and educational background?
- *Psychographics*: What is the psychological make-up of the target audience? Do they want the information to be provided in a direct way or do they prefer an engaging discussion format?
- *Attitude*: What are the employees' attitudes towards training and the learning content? What are their attitudes towards online learning?
- *Experience with online learning*: Will this be the employees' first exposure to using the company Intranet or Internet for learning or are they acquainted with navigating online materials?
- *Motivation*: What are the employees' work and career goals? How can the online course assist them to realise those goals?
- *Prior knowledge and experience*: What skills and knowledge relating to the subject will employees possess before training? To what extent do their skills and knowledge enable them to work towards achieving the desired business objectives?

c) Data analysis

After collecting data, the instructional designer completes a thorough analysis. The analysed data influences the outcomes of the design phase.

5.2.2   Design

In the design phase the outcomes formulated from data analysis are used to compile a blueprint for instruction. This blueprint is called a design and development blueprint (DDBP) and includes training needs, instructional strategies, content and assessment activities. The DDBP (referred to below as the blueprint) enables the instructional designer and project manager to communicate with members of the project/developmental team, and it keeps the project on track.

a) Determining learning outcomes

The first step in the design phase involves an examination of the tasks that were outlined in the analysis phase. From this the learning outcomes are formulated. The learning outcomes are specific and measurable.

b) Content outline

The learning outcomes determine the content of the learning program. The instructional designer develops the content outline after studying the data gathered in the analysis phase. The outline shows the breakdown of lesson-by-lesson topics and provides a motivational strategy that enables the program to attract and retain learner interest. The order of information in the outline moves from the known to unknown, concrete to abstract and easy to difficult.

c) Practical activities

The blueprint specifies the strategies for practical activities. Learners should be able to apply knowledge and skills during the learning process. Learners are provided with feedback regularly. Although the specific learning activities have not yet been created in this phase, a general description of practical activities is provided. The blueprint includes a brief description of assessment activities and workplace tasks.

d) Specifying mode(s) of delivery and media

In CIE, the choice of mode of delivery and media is influenced by the availability of computers, bandwidth, Internet access, software costs, plug-ins, the level of learners and learning centres. The blueprint identifies the technology. The company's information technology (IT) department then approves this technology. Technical specifications sometimes determine the choice of development tools. These include authoring tools, databases and learning management systems. The choice of delivery mode is based on instructional and technical standpoints.

The blueprint specifies how certain tools will be used in the learning program: text, graphics and images, audio and video, animation and 3D models, communication tools, electronic mail, mailing lists, discussion boards, chat tools, calendar tools, discussion tools, learner progress tools, passwords, assignment tools, learner grade tools and compiling tools.

The organisational profile is considered. This includes determining the expertise, infrastructure and readiness of the organisation to handle the chosen delivery mode and media. On the other hand, the delivery mode and media should meet the requirements of the subject and the needs of the targeted employees. The delivery mode accommodates the best practices of instructional theories: behaviourism, cognitivism and constructivism.

e) Determining the user interface

The graphical user interface serves as a link between learner and content. The blueprint indicates the buttons and navigational features that are used. The blueprint describes the look and feel of the learning program. The learner profile and the organisation culture determine the look of the visual treatment. In the corporate environment one deals with adult learners, so visual treatment may be relatively conservative, while the entire learning program is more creative.

f) Final sign off

Once the instructional designer has done the necessary revisions, the 'client' should regard the DDBP as a blueprint for the whole learning program. The client may be internal, such as the divisional manager or the project manager in the sponsoring department, or external. If you are an outside vendor, the client could be a company that is integrating CIE.

g) Rapid prototype

In the phase of rapid prototype, the learning software module is created for quick testing with a sample of learners, learning facilitators and subject-matter experts. The rapid prototype gives critical feedback on technical issues and the effectiveness of the learning program. The blueprint will then be modified to accommodate the feedback from the participants. The rapid prototype adds value to the effectiveness of CIE. The review process that is provided by the rapid prototype can identify errors and client preferences before the development phase. The technical issues could have a ripple effect on the entire program, so errors should be caught and corrected as early as possible.

South African organisations seem to have learnt from Kruse and Keil (2000:73), who urge instructional designers to ask the reviewers the following questions:
- Did the program immediately capture your attention?
- Was the creative theme engaging and appealing?
- Were the look and feel appealing? Was the program acceptable to corporate standards and culture?
- Was it easy to navigate the program? Did you ever feel lost or confused? Were the functions of the buttons easily identified?
- Was the quality of the audio acceptable?
- Was the quality of the video acceptable?
- Were waiting times acceptable during the loading and playback of graphics, animations and video?
- Was the tutorial lesson interactive and engaging?
- Did program features such as glossary, notepad and bookmark perform flawlessly?

5.3 DEVELOPMENT

Once the design and prototype have been completed, the team starts the development process:
- A storyboard is the first step in the creation of online learning materials. The storyboard provides a screen-by-screen description of what learners get in the instructional program.
- Graphic designers start to develop images that will be used in the learning program. Graphics may include illustrations, pictures and cartoons.
- When the graphic designer is developing the graphics, audio and video production can take place.

- Programmers then take the storyboard as a guide and develop the actual program.
- It will undergo an internal quality control. Experienced multimedia developers and naïve users who have no direct involvement in the learning program conduct the quality tests.
- Once the internal quality control has been finalised, the learning program goes through the formative evaluation/pilot test. A small group of people from the targeted learners and learning facilitators do the formative evaluation. The evaluation takes place in the learning environment and with the same computers that the rest of the learners will use. This evaluation enables the reviewers to check the technical bugs, glitches and effectiveness of instruction.

5.4 IMPLEMENTATION

The implementation is done when the computer-integrated program is ready for learning purposes. The program will be placed on the Internet or on a company Intranet. The author has observed that internal marketing of training programs to employees is an initiative that is usually overlooked. However, if the training program is compulsory, internal marketing is not necessary. Yet the organisation should not assume that employees will participate in training programs voluntarily without some form of marketing and encouragement.

Human resources development managers in South African organisations advise that a plan needs to be put in place to support learners and facilitators. End users should be assisted with technical issues. A help desk could be established to handle technical support calls and queries.

5.5 EVALUATION

The ADDIE model is an instructional design method that involves certain steps: analysis, design, development, implementation and evaluation. For the human resource development managers who participated in the study, evaluation is done in all phases of the ADDIE model. The author noted that the evaluation phase is usually ignored, however. This should be attributed to costs and additional time. On this premise one would advise organisations to employ Donald Kirkpatrick's model. As Kruse and Keil (2000:84) report, he introduced a four-level model of evaluation in 1975 that has become a classic in the corporate training environment:
- Level 1: Reaction
- Level 2: Learning
- Level 3: Behaviour
- Level 4: Results

These levels can also play a critical role in the CIE environment.

Level 1 Learners' reaction: Learners are to evaluate the effectiveness of training after completing the learning program. In most instances the evaluation forms that are filled in by learners measure how well they liked the training program. However, these forms will be more valuable if they go beyond this and ask complex questions. This evaluation is conducted in most organisations because it is easy and cheap to handle.

Level 2 Learning results: This evaluation will show whether learners have acquired the necessary knowledge, skills, values and attitudes that the program was supposed to instil and offer. To demonstrate achievement, learners should complete a pre-test and post-test that cover the relevant learning outcomes.

Level 3 Behaviour in the workplace: Learners usually perform well on the post-test, but the gist of the matter is whether they can retain and transfer the new knowledge and skills. In an ideal situation, the tests should be done three to six months after the training. The gap between training and measurement will allow learners to implement these skills and knowledge. Retention rates can also be evaluated. The learner, learner's supervisor and employees who report directly to the learner, and even the learner's clients, can complete behavioural scorecards.

Level 4 Business results: At this level the impact of the training program is evaluated. The performance of the trained employees is compared with the performance of employees who were not trained.

Kruse and Keil (2000:86) outline examples of the type of business impact data that can be evaluated:
- *Sales training:* Measurement of changes in sales volume, customer retention, length of sales cycle and profitability on each sale after the training program has been implemented
- *Technical training:* Measurement of reduction in calls to the help desk; reduced time to complete reports, forms or tasks; or improved use of software or systems
- *Quality training:* Measurement of reduction in number or severity of accidents
- *Management training:* Measurement of increase in engagement levels of direct reports

5.6 QUALITY ASSURANCE STRATEGIES

The research found that human resources development managers and facilitators in the corporate training environment used certain strategies:

a) Customisation

Customisation is employed by managers and facilitators to ensure that the content is of good quality: 'A lot of companies will buy generic content that is not customised.' This content is then customised to suit the South African environment. In most instances the content is of good quality, since it is a global product and the producer 'would ensure that the quality is fairly good'. Authors such as Gagné, Briggs and Wager (1992:7) and Kruse and Keil (2000:60) support this affirmation. The content is also usually aligned to international organisations or qualifications (such as the International Computer Driving License) which are held in very high esteem by the training industry.

b) Good instructional design as a strategy

Good instructional design strategy enables managers and facilitators to ensure that the content is of good quality: 'I would say content is good if it is instructionally well designed.' However, some facilitators and managers confessed that 'some of the content is not up to scratch; it's just paper that gets answered electronically and their wording's pretty bum'. Possibly some designers put more emphasis on multimedia than good instructional design: 'People often perceive that quality has to with rich media content – complex animations and interactiveness; a little more bells and whistles have been added to it.'

The majority of managers and facilitators in this study believed that 'it's all about sound instructional design'. They felt that the emphasis should therefore be on 'your audience analysis, your content analysis, the up-front design of the whole learning process, how it should take place, what should happen – where, what part of the material should really be developed by means of workshops?' These steps lead to the production of quality-assured content. The online content should be chunked well: 'You can't write a huge amount of content on there.' Managers had also put together instructional design teams to ensure that the content was of good quality: 'What happens then is that we have a design team that works with the subject-matter expert.' The involvement of the subject-matter expert was emphasised: 'We have a team of instructional designers; they meet the subject-matter experts.' Several authors support good instructional design as a sound instructional strategy (Elen and Lowyck 2000:252; Kwinn 2001:23; James 2001:16).

c) Legislative requirements

In the corporate training environment, managers and facilitators employ legislative requirements as a yardstick for content quality assurance: 'They draft the content up in such a way that it meets SAQA's requirements.' The content is also based on South African unit standards that are generated by standards generating bodies: 'If possible it is linked to unit standards.' There are many of these standards for organisations that participated in this study: 'The availability of standards is very good for most industries.' These standards are 'very outcomes-based and that content is handed over to an e-learning developer who actually converts it for whichever media is going to deliver it'. Despite that, some managers 'have taken the route not to align to NQF but rather to practise on their business competitiveness' because it takes considerable work to meet the legislative requirements. This problem is exacerbated because most managers 'don't know how to start' and 'they don't know how to end'. On the other hand, although the unit standards are available, they have not necessarily 'been constructed in an outcomes-based way' and they are not 'truly representative of all parts of industry – industries for that Seta'.

d) Research strategy

Managers and facilitators in this study employed a research strategy for quality assurance in online learning: 'We do research quite a lot.' Organisations usually conducted action and benchmarking research: 'Before we decided on one learning management system, we really did a lot of research into learning management systems.' The managers and facilitators acknowledged that 'putting research into practice is not easy' and for that reason 'they do peg their research a lot on what happens in America'. Some managers were of the opinion that South Africans do not 'do much research here'. Although research is done to improve practice and quality in the corporate environment, 'it happens in isolation; corporates should speak to each other more'.

e) Instructional design strategy: learners

Learners found that a good instructional design strategy was crucial for CIE because it enabled the course designers to do 'proper target analysis' so that 'you ensure that your target audience is right'. Learners believed that instructional design was important to the extent that 'every online facilitator needs the ability to design'. They also realised that sound instructional design would lead to outcomes-based learning materials. The development of learner-centred materials 'boils down to design principles'. They felt that 'the course has to be developed and designed to make sure you are reaching the outcomes, especially with outcomes; the outcomes have to be in accordance with the design'.

5.7 LEARNERS' PERSPECTIVE ON QUALITY ASSURANCE

During the fieldwork, it was found that learners perceived certain factors to be related to the quality assurance of computer-integrated content.

a) Content development teams

The researcher observed that managers put together computer-integrated content development teams consisting of various experts. These included the subject-matter expert, instructional designer, project manager, development team leader, web developer, interactive designer, critical reviewer, editor, graphic designer, electronic originator and quality assurer. The presence of these experts led to the design and development of quality content in line with the requirements of legislation.

b) Sound instructional design

Sound instructional design and development enabled organisations to produce online content of good quality. Employing a systematic and reflective process that took into account the principles of learning and training achieved this. These principles include conducting an instructional analysis, designing instructional and evaluation strategies and developing the learning materials according to the instructional strategy.

c) The requirements of legislation

The researcher observed that organisations tried to ensure that the computer-integrated materials complied with the principles and requirements of OBET, the NQF and SAQA. However, some paper-based learning materials that were being used in conjunction with the online content were not really outcomes-based, since they lacked outcomes, learning activities, feedback on activities, examples and graphic elements. The researcher also observed that off-the-shelf materials, which were mostly imported from other countries, did not necessarily satisfy the requirements of South African legislation. Be that as it may, there are endeavours within organisations to customise the off-the-shelf materials. A few organisations have started to produce local off-the-shelf content for soft skills courses. Some of these courses address local issues and challenges such HIV and Aids in the corporate environment.

5.8 COMPUTER-INTEGRATED LEARNING AND USER INTERFACE DESIGN

The user interface establishes the graphical associations of the learning program in the minds of the facilitator and learners (Kruse and Keil (2000:107). In fact, some users see the user interface as a learning program in the same way that a well-designed book cover becomes the reader's visual image of the quality of its content. This section will address the factors that are critical in the user interface: introduction of a program, learner control, presentation of information, providing help and ending a program.

5.8.1 Introduction of a program

Instructional designers should make sure that the learning program has a title page (Lynch and Horton 1999). According to Alessi and Trollip (2001:48), a title page serves certain purposes:
• To tell the user which programs she/he is about to use
• To tell the user in a general way what the programs are about
• To attract user attention and create a receptive attitude
• To inform the user of the author's or publisher's name and contact information
• To provide copyright information
• To provide an escape for the user if she/he has come to the wrong place

The title page should contain elements to motivate learners. However, Alessi and Trollip (2001:48) warn that these motivational elements should not make the title page unnecessarily long. The introduction of the learning program should also have directions. Directions should match the level of the learners (Lynch and Horton 1999). Lack of directions frustrates learners and, according to Kruse and Keil (2000), learners express the following complaints:
• 'What am I supposed to do now?' This frustration is a consequence of poor or deficient directions.
• 'Did I finish everything?' This dissatisfaction is a result of too many hyperlinks and layers of content. Developers should provide learners with a path of navigation so that they are not lost in hyperspace.

- 'What's it doing? Is it hung up?' This complaint is a result of the slowness of computers if they are processing large programs. Learners' anxiety can be eliminated by providing learners with a computerised message, for example, 'Loading program, please be patient.' Directions should be simple.

South African CIE developers indicated that it is advisable to include the user identification element in the learning program. The rationale is to store/retrieve information and make sure the user is authorised to access the program. User identification also enables the program to address the user by his/her name throughout, making it learner-centred and user-friendly.

5.8.2   Learner in control

An effective user interface puts the learner in control of the learning program. The programmers and instructional designers reasoned that when learners are given control, their anxieties and frustrations are reduced, and thus learning becomes effective. This view is supported by Alessi and Trollip (2001:52), who write that learner control should include the following:
- The most important learner control concerns sequence, which includes moving forward, moving backward, and selecting what to do next. Learners should also be able to control the pace of learning. More optional controls are for content difficulty and learning strategy.
- Always allow learner control of forward progression.
- The developer should not include timed pauses. A timed pause is a progression to the next step after a fixed number of seconds, without requiring the user to click or to press a key.
- Allow the learner to review whenever possible, such as with backward paging.
- Always allow the learner temporary termination of a program. This means being able to end the program for the time being and return later.
- For general capabilities, such as directions, help, complaints, glossaries and temporary termination, learners should be provided with consistent global control everywhere in the program.
- Whenever there are movies, audio or animations, learners should be allowed to pause, continue, repeat or skip them. If the movie or other information is long, fast-forward and rewind controls should be provided.
- Learners with more prior knowledge of the content should be given greater control than those with little content experience.
- The developer should know his/her users so that controls can be appropriate to their needs.
- Learner controls should be based on content. Generally, more learner controls should be provided for problem solving and high-order thinking skills and more program control for procedural learning and simpler skills.
- If the mastery of content is critical, more program control should be provided.

Learners used various tools for control. These included the mouse and keyboard. The mouse should be used as a primary mode of control and the keyboard as a secondary mode. The developers may include less popular interfaces such as screens and voice-recognition systems that give learners more control. Learners also used buttons and menus (for example full-screen menus, hidden menus and frame menus).

5.8.3 Presentation of information

The way in which information is presented has a serious impact on the effectiveness of the learning program. Presentation includes issues such as the appearance of what is displayed on the screen, visual elements, layout, text legibility, colour, graphics and multimedia components such as audio, video and interactive elements, help and ending the program.

5.8.4 Consistency

According to the program developers and instructional designers in South African organisations, the methods of presenting information in learning programs should be consistent. The most efficient way to achieve consistency is to divide the screen into functional areas.

5.8.5 Modes of presentation

Modes of presentation include text, graphics, sound and video (see figure 5.1).

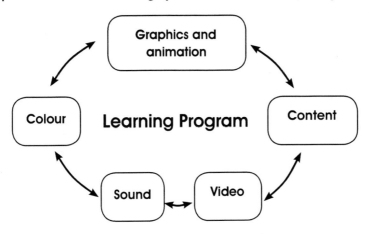

FIGURE 5.1 MODES OF PRESENTATION

Figure 5.1 illustrates how various modes of presentation can be integrated in a single learning program. However, it does not necessarily mean that each learning program should have all these modes. The selection and integration of modes will be done after learning and resources analysis has been carried out.

a) Text information

Generally, the South African developers followed guidelines for paragraphs of regular text as outlined by Alessi and Trollip (2001:61):
- *Adequate type size:* Usually, the developer should use a 10-point or larger font. In HTML documents, only type size 2 and larger can be used.

- *High background contrast:* Make sure that the text stands out from the background. The developer should avoid fussy backgrounds and those that are similar in colour and brightness to the text (see figure 5.2).

FIGURE 5.2 FUSSY BACKGROUND

It is difficult to read the content on the 'computer screen' (figure 5.2), because the background is too similar in colour and brightness to the text.

- *Left-aligned text:* The developer should avoid centred or right-aligned text except for short passages and then only for a special effect (see figure 5.3).

FIGURE 5.3 CENTRED AND RIGHT-ALIGNED TEXT

Although the background in the 'screen' in figure 5.3 is not similar in colour and brightness to the text, the content is difficult to read because the paragraphs are centred or right-aligned.

- *Short line length:* The developer should keep lines of text to 40–60 characters.
- *Adequate line spacing:* Space lines of text should be about 1/30 the length of a line.
- *Uppercase and lowercase:* The developer should not use ALL UPPERCASE for an entire paragraph (see figure 5.4).

FIGURE 5.4 USING UPPER CASE FOR ENTIRE PARAGRAPH

It is very tiring to read the paragraphs on the 'screen' (figure 5.4) because uppercase is used for the entire page. Capitalised text is a less effective method of typographical emphasis. To read a paragraph of text in uppercase letters, the reader must parse the letter groups – read the text letter by letter – which is very strenuous.

- *Simple character shape:* Fonts with simple letter shapes are preferable. Developers of learning programs should avoid decorative and highly stylised fonts. Fonts such as Arial and Tahoma work well for a screen display. When the learning program includes long passages of plain text, the instructional designer should consider how many learners will have the stamina to read and understand the block. Kruse and Keil (2000:115) support this view. The instructional designer should also bear in mind that learners read 25 per cent more slowly from a computer screen than from paper (Kruse and Keil 2000:115). Using a computer-integrated program with paper-based material or blended learning could solve this.

b) Graphics and animation

The program developers and instructional designers claimed that wise use of graphics and animations adds a lot of value in learning. However, if graphics and animations are wrongly used, they are detrimental to the learning process. Graphics should be designed for important information; otherwise they distract learners unnecessarily. In designing and using graphics, the graphic designer and instructional designer should consider the purpose of graphical information and the graphic type.

There are four primary uses of graphics during the presentation of the learning program (Alessi and Trollip 2001:68):
- As primary information
- As analogies or mnemonics
- As organisers
- As cues

The main task of graphic design is to create a strong, consistent visual hierarchy where significant content is emphasised and elements are well organised. Alessi and Trollip (2001:71) outline the main types of graphical information:
- Simple line drawings
- Schematics
- Artistic drawings
- Diagrams
- Photographs
- Three-dimensional images
- Animated images

In the corporate training environment, graphic and instructional designers have to be familiar with all types of graphics so that they can design and choose the appropriate one for a particular purpose.

c) Video presentations

Video presentations combine visual and auditory information, which can be very valuable in a learning program. Videos were used to show learners how to do certain things with a recording of the live activity (simulation). The ability to combine visual simulation and speech in a video can add a lot of value to the learning process. Unfortunately, Alessi and Trollip (2001:63) warn that if video is improperly used, it may have negative effects on the learning process. For example, reading text and hearing speech simultaneously may decrease learning. On the other hand, if the text being read by the learner is exactly the same as what is being spoken, the consequences can be worse. Instructional designers who participated in this study suggested that speech should be used with brief bulleted points or to describe graphics and diagrams.

Video has become a common component in interactive multimedia (see for example McCormack and Jones 1998:107). It may take many forms, such as a soundless demonstration of a procedure, a narrator describing a live activity, cartoons, interviews and play (Alessi and Trollip 2001:72). When the real actions cannot be captured, realistic animation can be used.

The use of video in learning and training has opened many opportunities and advantages for South African organisations. Alessi and Trollip (2001:72) outline some of them:

- Dramatisations encourage learners to evaluate their attitudes and thus are useful to effect attitude change.
- It is more effective to listen to a narrator explaining how to operate a complicated device than to read a text passage about it.
- Video can be engaging, entertaining and thought provoking.
- Video can lend a very professional look to a program.
- A short video can replace long convoluted text passages.
- Video is a natural choice for some learning activities, such as conversation in a foreign language or analysis of events in social studies.
- Learners may find that information presented by video is more memorable because of its visual detail and its emotional impact.

Although video can add a lot of value to learning, designers were aware of its costs and pitfalls. According to Alessi and Trollip (2001:73), producing a quality video can cost thousands of American dollars per minute. On the other hand, if a poor video is produced in an effort to cut costs, the professional look of the learning program can be affected negatively.

d) Sound

According to developers in the corporate training environment, sound, especially speech, is very important in a CIE program. It also has its advantages and pitfalls. Despite that, sound is very important for some content areas. Sound is also crucial when it comes to gaining learners' attention.

Suggestions for the proper use of sound (audio) are outlined by Alessi and Trollip (2001:75):

- Use speech for gaining attention, directions and dual coding.
- Provide speech for users who have difficulty in reading text.
- Provide both text and speech as options.
- Use audio for appropriate content areas, such as language learning.
- Allow user control of audio (pause, continue, repeat, skip and volume).
- Allow the usual program global controls, even during audio segments.
- Do not use token audio in just one or two places.
- All inserts must be of high quality.

CIE can be greatly enhanced with sound if it is properly incorporated.

e) Colour

The use of colour is similar to that of graphics. Colour is very effective in the learning program and may be used freely, if judiciously.

Colour enables the learning program to attract the attention of learners.

Suggestions for the proper use of colour are provided by Alessi and Trollip (2001:77):
- Use colour for emphasis and for indicating differences.
- Ensure good contrast between foreground and background colours, especially for text.
- Use only a few colours for colour coding.
- Use colours in accordance with social conventions.
- Be consistent in the use of colour.
- Test programs on non-colour displays to assess their effect on persons with colour vision deficiency or with older equipment.
- Balance learner affect and learning effectiveness when using colour.

5.8.6 Providing help to learners

The developers ensured that learners always obtained help, including procedural help. Informational help should be provided to assist learners with content.

Alessi and Trollip (2000:77) make certain recommendations for helping the user:
- Always provide procedural help.
- Provide informational help, depending on the program's purpose and methodology.
- When providing informational help, try to make it specific to the content.
- Allow a return to directions at all times.
- Provide help via rollovers for functions at any particular time.
- Always have a help button or menu visible, to remind learners that help is available.
- Provide help for starting a program in a print manual. Online help is not useful if the user does not know how to start the program.

5.8.7 Ending the program

The learner may end a program temporarily or permanently. Alessi and Trollip (2001:81) provide recommendations for closing and exiting the learning program:
- Provide the facility for the user to exit anywhere in a program.
- Ensure that a temporary exit is always available with user control.
- Provide a safety net to rescind a request to exit.
- Provide closing credits with control.
- Provide a final message that makes it clear the user is leaving the program.
- Return the user to an appropriate place after the program quits.

5.9 EVALUATION INSTRUMENTS FOR COMPUTER-INTEGRATED COURSES

During the analysis of documents collected from corporate South Africa, the researcher could not come across an evaluation instrument for determining the quality of courses, especially those that are computer-integrated. The only documented quality instruments were obtained from the higher education institutions. However, towards the end of the study in 2004, some organisations and vendors were starting to introduce and implement these instruments. An example from a higher education institution is provided below.

If you are designing and writing online learning materials, you should find this checklist a useful evaluation tool. The items are clustered under specific headings representing core elements of OBE and open learning materials. You should critically analyse your materials by indicating whether the core elements are adequately presented/displayed and applied/implemented materials. Please tick the appropriate box.

TABLE 5.1 CHECKLIST FOR WELL-DESIGNED ONLINE LEARNING MATERIAL

	CORE ELEMENTS	YES	NO
A	**PREFACE**		
1	Has the approach to outcomes-based education in this particular program been explained to the learner?		
2	Have you given an indication of the time (notional hours or credits) the learner should spend on achieving the outcomes of the module/unit?		
3	Does the preface alert the learner to the way the materials (learning package) are designed to work/explanation of the different aspects of the materials/package?		
4	Does the preface give a short summary of the core elements and purpose of the module?		
·5	Is there a clear indication of any prerequisite knowledge or skills?		
6	Are the benefits of the learning program indicated for the learners?		
7	Are the icons used in the learning guide explained?		
B	**OVERVIEW and/or INTRODUCTION**		
8	Is there an overview of the whole unit in the event of one unit consisting of more than one part/section?		
9	Is the bigger picture given to the learner, in other words, what exactly is going to be covered, how is it going to be dealt and why is it dealt with in this way?		
10	Is each part introduced in an interesting, stimulating way?		
11	Is the learner oriented concerning the unit in terms of the module or learning program?		
12	Is there a visual representation (diagram) to show where the units fit into the whole picture of the program?		
13	Does the learner know where the theme fits into the work environment?		
14	Is there reference to learning objectives and how they integrate with the unit's activities and content?		
15	Does the learner have an indication of the benefits of achieving the learning objectives?		

16	Does the learner have a clear indication of the responsibilities of monitoring his/her own progress in each unit and what he/she must look for?		
C	**OUTCOMES AND LEARNING OBJECTIVES**		
17	Do the learning outcomes and objectives directly correspond with the overall outcomes of the program?		
18	Do the learning outcomes and objectives reflect what the learner must be able to do in the work situation?		
19	Are the outcomes and objectives correctly and clearly expressed (SMART: specific, measurable, achievable and realistic in terms of time)?		
20	Are the learning outcomes and objectives designed according to the relevant level descriptors of the NQF framework/stated qualification?		
21	Do the learning outcomes and objectives and, by implication, the learning guide focus more on applied competence than on content or theory/knowledge?		
22	Do the learning outcomes mirror the end result of learning, in other words, the integration of the applicable knowledge, skills, values and attitudes?		
23	Do the learning outcomes and objectives link (match) with the learning experiences/activities and applied assessment methods?		
D	**INTERACTIVE, LEARNER-FRIENDLY TEXT**		
24	Is there evidence that you have considered the learner profile and needs when compiling the learner guide, i.e. are the language, style and all elements of the learning package appropriate for the target audience?		
25	Is a conversational and personal style used in the learning guide, e.g. use of the first person?		
26	Is the vocabulary accessible to all learners? Is the readability level of the text appropriate for the learners?		
27	Are the language, activities and examples used sensitive to different cultures, gender and disabilities?		
28	Does the tone of the language vary, e.g. sympathetic/humorous/challenging?		
29	Does the construction of the sentences help to maintain logical flow?		
30	Are the sentences constructed in the active voice rather than the passive voice?		
31	Have you used shorter sentences and paragraphs to enhance effective chunking of information into sections/units/modules?		
32	Has the learning content been organised into logical, manageable units or sections according to the outcomes?		
33	Is the numbering system straightforward and easy to follow (preferably no more than three levels in the hierarchy)?		

34	Does the text build continuously on prior knowledge?		
35	Are the concepts, facts, examples and opinions clearly identified and defined?		
36	Are there enough realistic and functional examples of concepts?		
37	Is there good signposting (prompts) among components of the learning guide (relevant icons and/or directions to different elements and media)?		
38	Have you used different techniques to motivate learners to master applicable knowledge, skills and attitudes?		
39	Is enough information provided to the learner to achieve the desired outcomes?		
E	**LEARNING EXPERIENCE/ACTIVITIES, FEEDBACK AND ASSESSMENT (FORMATIVE AND SUMMATIVE)**		
40	Are the learners continuously and actively exposed to different integrated learning processes (cognitive, affective and psychomotor)?		
41	Are the learning experiences personalised, e.g. related to workplace situations?		
42	Is there a variety of activities to accommodate different learning styles?		
43	Do the learning strategies address the needs of the learners in terms of applicable theories?		
44	Are all the activities focused on mastering the learning outcomes and objectives?		
45	Are the learning activities integrated and prominent in the learning guide, and do they serve as formative assessment activities?		
46	Are the instructions set in the activities or questions clear for the learners?		
47	Are the activities inviting and challenging – is it clear to learners that it is valuable for them to complete rather than skip the activities?		
48	Is there enough space for the learners to write their answers in case of the learning guide being used as a workbook?		
49	Collectively, do the activities and assessment questions (formative and summative) guide and focus on the learner's achievements of the outcomes/objectives?		
50	Have appropriate forms of assessment been identified, explained and integrated (e.g. self-assessment, portfolios, etc.)?		
51	Have learners been directed to other sources to integrate knowledge, skills, values and attitudes into their field of learning?		
52	Have learners been encouraged to integrate knowledge, skills, values and attitudes from other fields to achieve an outcome (where applicable)?		
53	Are the criteria for assessment of competence clearly indicated?		
54	Do the criteria for learner competence correlate with the overall assessment criteria for the module?		

55	Are learners encouraged to apply critical thinking and problem-solving skills during learning and to attempt the learning activities/assignments?		
56	Are learners encouraged to reflect on their learning?		
57	Do learning assignments reflect the achievement of outcomes?		
58	Have you included feedback or guidelines for the learners on the learning activities and formative assessment activities (where applicable and necessary)?		
F	**STRUCTURE, LAYOUT, MEDIA AND SUPPORT**		
59	Is the material visually attractive, with functional page design?		
60	Are there functional and labelled tables, diagrams, lists, flow diagrams, mind maps and good quality graphics between the text?		
61	Is there sufficient white space for learners to write notes, answer questions, do calculations, etc.?		
62	Are all media integrated with and relevant to other components?		
63	Is the whole course (text and other media) well designed and can the learner easily navigate the course?		
64	Are methods and sources of learner support built into the learner guide?		
65	Is information provided on portability, progression and articulation possibilities into other related learning areas?		
66	Does the learning package cover the critical outcomes within the framework of the learning program to develop people to become self-regulated and directed learners while developing lifelong learning skills?		
67	Does the total volume of the learning package correspond with the number of credit points that a learner will get after achieving the outcomes of the unit/module?		
G	**SUMMARY/CONCLUSION**		
68	Is there a clear and useful summary/conclusion at the end of sections or units that reflects objectives?		
H	**GENERAL/MISCELLANEOUS**		
69	Have learners and other stakeholders been requested to provide feedback on the quality of the learning guide?		
70	Is a complete contents list with corresponding web references (URL) provided?		
71	Are all required and optional resources listed?		
72	Is an adequate glossary provided or is the relevant terminology explained where necessary?		
73	Is a complete bibliography or reference list included at the end of the unit/module?		
74	Have you adhered to copyright regulations?		

Source: Adapted from The CCDD Quality Assurance Framework (Technikon SA) (2003)

5.10 CRITERIA FOR QUALITY COURSEWARE

The criteria for quality courseware were very useful in the provision of the quality courseware.

TABLE 5.2 CRITERIA FOR EVALUATING QUALITY COURSEWARE

Tick the most appropriate box		
	YES	NO
1 **Orientation to program, introductions, aims and learning outcomes**		
1.1 **Introductions to program/modules/sections**		
– Explain the importance of the topic for the learner, and create interest in the material – Provide an overview of what is to come – Forge links with what the learners already know and what they are expected to learn – Point out links with other lessons/sections – Provide some indication of intended learning outcomes in ways that are directly relevant and useful to learners – Give indications of how much time the learner should spend on the material in the lesson so that learners can pace themselves		
1.2 **Learning outcomes**		
– Are stated clearly and unambiguously – Describe what learners need to demonstrate in order to show their competence – Are consistent with the aims of the course and program - The content and teaching approach support learners in achieving the learning outcomes		
2 **Selection and coherence of content**		
– Content is current and reflects contemporary thinking and recent references – Content is appropriate both to the intended outcomes of the program and to recognising prior learning – Content builds on learners' experience where possible – There is appropriate variety in the selection of content – The content and teaching approach support learners in achieving the learning outcomes		
3 **Presentation of content**		
– Concepts develop logically – Concepts are explained clearly using sufficient and relevant examples – New concepts are introduced by linking to learners' existing knowledge – Ideas are presented in manageable chunks – A variety of methods is used to present the content and succeed in keeping the learners' interest alive – Theories are not presented as absolute – debate is encouraged – The course materials model the processes and skills that learners are required to master: that is, they practise what they preach		

4	View of knowledge and use of learners' experience		
	– Learners' own experience and understanding are seen as valid departure points for discussion – Knowledge is presented as changing and debatable rather than as fixed and not to be questioned – Learners are encouraged to weigh ideas against their own knowledge appropriately, and a concerted effort is made to empower learners to use theory to inform practice		
5	Activities, feedback and assessment		
5.1	Activities		
	– The activities are clearly signposted, and learners know where each begins and ends – Clear instructions help learners to know exactly what they are expected to do – Activities are related to the learning outcomes – Activities reflect effective processes – Activities are sufficient to give learners enough practice – Activities are distributed at fairly frequent intervals throughout a section – Activities show a range of difficulty – Activities are sufficiently varied in terms of task and purpose – Activities are life/work–related – Activities are realistic in terms of time indications and resources available to learners		
5.2	Feedback to learners		
	– Feedback to learners is clearly indicated – Feedback is offered in the form of suggestions and is prescriptive only where necessary – Learners are able to identify the errors they have made, and they are to assess their progress from responses – Where calculations are required, stages in the working are displayed and explained		
5.3	Assessment		
	– There is an assessment strategy for the course as a whole – The assessment tasks are directly related to the learning outcomes – Formative and summative assessment strategies are employed – Assessment criteria are made known to learners, and feedback is provided on interim assessments, which helps learners to improve – Mechanisms exist for learners to respond to feedback on assessment, and these are clearly explained in the courseware		
6	Language level		
	– New concepts and terms are explained simply, and these explanations are indicated clearly in the text – The language used is friendly, informal and welcoming – Learners are not patronised or 'talked down to' – The course is appropriate to the learning intended – The language is sensitive as far as gender and culture are concerned – The language takes cognisance of the multilingual reality of South Africa – The language is active and sufficiently interactive		

7	Layout and accessibility		
7.1	Learning skills		
	– Summaries and revision exercises are included at a frequent interval to assist learners to learn – Skills for learning (such as reading, writing, analysis, planning, managing time, evaluation of own learning needs and progress) are appropriate to the outcomes of the course, and integrated into the materials		
7.2	Access devices		
	– The numbering/heading system makes it easy for learners to find their way through the text – The text is broken up into reasonable units – Headings and sub–headings are used to draw attention to the key points of the lesson. This makes it easy for the learners to get an overview of the lesson at a glance. It also makes it easy to find the parts that learners want to refer to – There is a contents page – Pre–tests are used wherever feasible to help learners know what skills or knowledge they need to have before starting the lesson/ section – Links with previous knowledge and experience, with other parts of the same lesson, and with other lessons are indicated		
7.3	Visual aids in text		
	– The visual aids used complement the written text – Line pictures and cartoons are well drawn and appropriate for the target learners. They are gender and culturally sensitive – Where appropriate, concept maps and diagrams are included to help learners get an overview of the material and to assist the learning process – Captions and explanations accompanying visual aids are adequate and give learners a clear idea of their purpose		

Source: Adapted from Technikon South Africa and South African Institute for Distance Education (2002). Evaluation Capacity-Building Project: Report Part 3 – Evaluation Instruments

5.11 RECOMMENDATIONS CONCERNING THE USE OF QUALITY ASSURANCE STRATEGIES

- A lot of emphasis should be put on good instructional design as a strategy for quality assurance purposes in CIE. This will prevent facilitators from dumping chunks of text online – something which does not satisfy the requirements of OBET.
- Managers and facilitators should attend workshops on learning materials development, generation of unit standards, satisfaction of SAQA requirements and customisation of content. The findings of this study indicate that most managers do not know where to start with regard to these aspects.

- Organisations should put more resources into research. Most research activities conducted into CIE are very elementary. Organisations in the corporate training environment should cooperate with South African and other higher education institutions in online learning research.
- All Setas should be involved in assisting organisations within their industrial sectors to transform human resources development through the integration of CIE. During this study only the Bankseta was engaged in assisting organisations in the banking sector with the integration of CIE. The Bankseta was providing many resources for CIE. Setas that are not doing this are not fulfilling the mandate they have been given by legislation.
- The full-blown integration of CIE should be preceded by the formulation of a business case and strategy for CIE as well as a pilot project. This will increase the buy-in of stakeholders.

CHAPTER 6

COMPUTER-INTEGRATED ASSESSMENT: CONCEPTS AND PHILOSOPHICAL UNDERPINNINGS

CONTENTS

6.1   INTRODUCTION

This chapter, which reflects on computer-integrated assessment (CIA) concepts and philosophical underpinnings derived from the corporate training environment, aims to assist readers in comprehending, developing and executing assessment practices that are suitable for CIE, training and development. CIA in the corporate training environment is a structured process of accumulating valid and reliable information regarding the performance of the learner. This is continuous, based on well-defined criteria, using various methods, computer-integrated tools and strategies, noting the results, and providing necessary feedback and support to the learner.

The feedback may also be provided to the facilitators, assessors, human resources development managers, line managers, learning designers, subject experts, project managers, development team leaders, critical reviewers, web designers and other relevant role players and stakeholders. CIA may be viewed as a way of passing a judgement on learner competence in a computer-integrated learning environment. This judgement should be based on learning outcomes and organisational, national and international standards. CIA, in other words, should measure knowledge, skills, values and attitudes. The data collected enables the assessor to evaluate the learner performance, learning materials, teaching and learning methods. The identified learner's achievement should be interpreted and recorded. This information can then be used to improve the teaching and learning process.

6.2 PURPOSE OF COMPUTER-INTEGRATED ASSESSMENT

The purpose of employing CIA is to assess learners with the intention of supporting them and improving their understanding and performance. CIA should be used to detect the learner's progress in a particular area of learning. This enables the facilitators and learning designers to take appropriate action regarding the manner in which further teaching and learning can be facilitated, based on the learning outcomes. Of course the learning outcomes would include knowledge, skills, attitudes and values.

CIA reveals information about learning problems and possible remedial action for learners who are experiencing difficulties. While the aim of CIA has traditionally been to promote or fail learners, the corporate learning environment emphasises learner progression.

In the South African context, the NQF requires CIA to:
• Determine whether the learning required for competence in specific outcomes is occurring and whether any difficulties are being addressed
• Report to the learner and/or any other role players and stakeholders the rate at which competence is acquired during the learning process and to construct a profile of the learner's competence across the curriculum
• Identify information for the evaluation and reconsideration of the learning programs used in CIE
• Optimise the learner's accessibility to the knowledge, skills, attitudes and values as defined in the organisation's human resources development policy.

6.3 PRINCIPLES

According to CETA (2001), several principles underlie CIA:
- The competence of a learner is established against the assessment criteria of a unit standard using the assessment criteria and statements of competence.
- Each of the presented assessment elements (assessment interview, assessment assignment and debrief interview) has a separate function in establishing the competence of the learner.
- The CIA should ensure that all the specific outcomes, critical cross-field outcomes and essential embedded knowledge are assessed.
- The specific outcomes must be assessed through observation of performance. Supporting evidence should be used to prove competence in specific outcomes only when they are not clearly seen through actual performance.
- Essential embedded knowledge must be assessed in its own right, through oral and/or written evidence. It cannot be assessed only through seeing knowledge being applied.
- The specific outcomes and essential embedded knowledge must be assessed in relation to each other. If a learner is able to explain the essential embedded knowledge, but is not able to perform the specific outcomes, he/she should not be assessed as competent. Similarly, if a candidate is able to perform the specific outcomes, but is unable to explain or justify his or her performance in terms of essential embedded knowledge, he/she should not be assessed as competent.
- Evidence of the critical cross-field outcomes should be found in the performance and in the essential embedded knowledge.
- Performance of specific outcomes must actively affirm target groups of learners. Learners should be able to justify their performance in terms of their values.

Given these principles for establishing the competence of a learner, competence is measured in several ways. CETA (2001) goes further and reports that in order to ensure achievement in how CIA is delivered, the principles reflected in table 6.1 should be adhered to:

TABLE 6.1 PRINCIPLES OF ASSESSMENT

Principle	Definition	Majority experience
Integration	Forms part of human resources development, which provides for the establishment of a unifying approach to education	Separation by race, sex, ages, academic and technical or vocational education. Education and training were front-end experiences. Older people were regarded as learners
Relevance	Being and remaining responsive and appropriate to national development needs	Little match between what was taught in school and what was required to be employed or enter further/higher learning

Credibility	Having national and international value and acceptance	Certificates and qualifications from some providers and institutions enabled a few to access national and internal system
Coherence	Working within a consistent framework of principles and certification	No or little linkage between the education and training parts of the system. No means (except time) to measure difference in learning across providers or courses (programs)
Flexibility	Allowing for multiple pathways to the same learning ends	At higher levels, access restricted by professional qualifications and certificates within narrowly defined occupational 'fields'. Little or no room was made for learning through methods other than institution-based instruction
Standards	Being expressed in terms of a nationally agreed framework and internationally accepted standards	Sector, enterprise, province, NGO, state, institution – all had their own particular, unique requirements for learning achievements
Legitimacy	Providing for the participation of all stakeholders in the planning and coordination of standards and qualifications	Little or no consultation; use of experts mainly. No cooperation or consultation between the DoE and the DoL
Access	Providing ease of entry to appropriate levels of education and training for all prospective learners in a manner that facilitates progression	Entry by certificate based on years of study with few mechanisms for alternative entry. Entry often restricted by requirements based on race, age and sex, not outcomes or learning requirements
Articulation	Providing for learners, on successful completion of accredited prerequisites, to move between components of the delivery system	Standards and entry requirements for education and training programmes set at institutional and provider level with little national similarity by institution, faculty, programme or company
Progression	Ensuring that the framework for qualifications permits individuals to move through the levels of national qualifications via appropriate combinations of the components of the delivery system	Rather than stepping through a clearly sequenced series of education and training requirements for further and higher levels in a pathway, learning was negotiated through a series of glass ceilings where entry requirements at the level above were markedly not the same as exit requirements for the level at which learners found themselves

Portability	Enabling learners to transfer their credits or qualifications from one learning institution to another	Training generally was sector- or employer-specific, locking learners in because there was no system of ensuring transfer across organisations, sectors, employers or even provinces. Certification based largely on what learners could do or say, not on the integration of knowledge, understanding, skills, context, and abilities such as communication, problem identification, problem solving, reflection and application to other situations and contexts
Recognition of prior learning	Through assessment, giving credit to that which has already been acquired in different ways	The education-while-young system meant that those who had not achieved particular levels of literacy in the official languages were generally regarded as having stopped learning at whatever educational level they had achieved and had not learnt anything from social or work interactions. Older people were not encouraged to re-enter the learning environment
Guidance of learners	Counselling of learners by specially trained individuals who meet nationally recognised standards for educators and trainers	While the previous education and manpower structures had different vocational and careers guidance programmes, these were separate and out of touch with the learning and employment needs of the majority of people and/or industry, economic sectors and development needs

Adherence to these principles in CIA can contribute to creating a sound relationship between work and learning, between learning and applied competence, between learning and employability, and between qualification and employability (CETA 2001). The principles promote quality and equity in the workplace learning environment. They also ensure that there is an effective and valuable education, training and development system. This kind of system promotes equal benefits for all South Africans.

6.4 COMPUTER-INTEGRATED ASSESSMENT VERSUS CRITICAL OUTCOMES

The South African Qualifications Authority Act (1995) contains a list of seven critical outcomes that are regarded as being important to skills development in all learning areas. These critical outcomes influence the way in which CIA will be planned and implemented. In other words, the assessor should take them into consideration when developing CIA. In fact, computer-integrated learning material should integrate these outcomes. At the same time, corporate experience has shown that it is not always easy to integrate them meaningfully into unit standards.

The following critical outcomes should therefore be integrated into CIE programs and assessment:
- Identify and solve problems in which responses display that responsible decisions using critical and creative thinking have been made.
- Work effectively with others as a member of a team, group, organisation and community.
- Collect, analyse, organise and critically evaluate information.
- Communicate effectively using visual, mathematical and/or language skills in the modes of oral/or written presentation.
- Use science and technology effectively and critically showing responsibility towards the environment and health of others.
- Demonstrate an understanding of the world as a set of related systems by recognising that problem-solving contexts do not exist in isolation.

To maximise the full personal, professional, social, intellectual and economic development of each learner, organisation and society at large, the intention underpinning any computer-integrated learning program and assessment should make an individual learner aware of the importance of:
- Reflecting on and exploring a variety of strategies to learn more effectively
- Participating as responsible citizens in the life of local, national and global communities
- Being culturally and aesthetically sensitive across a range of social contexts
- Exploring education and career opportunities
- Developing entrepreneurial abilities

6.5 COMPUTER-INTEGRATED ASSESSMENT IN OUTCOMES-BASED EDUCATION AND TRAINING

The *Government Gazette* (South Africa 1998) and Gauteng Department of Education (GDE) (1999) claim that computer-integrated assessment in OBET assists learners to reach their full potential; and computer-integrated assessment in OBET is participative, self-referencing, criterion-referenced, involves a move away from rote learning and entails learners using knowledge in real-life situations. Other sources confirm what the author has noted in the corporate training environment, namely that there is a paradigm shift in the approach to CIA. CIA in OBET should focus on the critical and specific outcomes, be guided by learning activities and criteria, ensure that outcomes are valid, and cater for learners with disabilities.

6.5.1 Assisting learners reach their full potential

OBET is learner-centred and outcomes-based, which could enable CIE to consider that all learners must and can achieve their maximum potential (although this need not take place uniformly or within the same timespan).

CIA in OBET should focus on the attainment of well-defined outcomes, creating the possibility of crediting learners' achievements at every level, whatever pathway they pursue, and at whichever rate they may have gained the necessary competencies.

CIA should use tools that can appropriately assess learner competence and encourage lifelong learning in the working environment. CIA in OBET is regarded as the best model to assess outcomes of learning through the organisational system and enable improvements to be effected in CIE. Assessment should support the learner developmentally and channel feedback into CIE, training and development. This means that CIA should not be seen merely as the collection of a string of traditional test results.

6.5.2 Participative
Computer-integrated OBET is based in the notion of the participation of other role players and stakeholders in CIE, training and development to ensure transparency. These include facilitators, assessors, human resources development managers, line managers, learning designers, project managers, development team leaders, critical reviewers, web designers and other relevant role players and stakeholders who may assist in the solution of learning problems or in supporting and encouraging learners. CIA in OBET also promotes transparency by presenting the critical and specific outcomes clearly.

6.5.3 Self-referencing
Self-referencing in CIA can also be very helpful. The GDE (1999) refers to self-referencing as ipsative assessment, which compares a learner's achievement with that of other learners who are participating in CIE.

6.5.4 Criterion-referenced
CIA in OBET should be criterion-referenced. Assessors and learners should share the criteria for CIA. They should also assess performance with reference to the agreed criteria. Facilitators and peers should provide feedback so that learners' self-assessment skills are developed. In other words, learners are empowered to be in charge of their learning development.

6.5.5 A move away from rote learning
The development ethos of CIA should largely be influenced by constructivism, a paradigm that claims learners should construct their knowledge. Constructivism discourages the notion of facilitators being perceived as a source of knowledge and learners as merely passive receptors. Constructivism promotes the application of the acquired skills and knowledge, so learners do not have to resort to memorisation. This has an impact on the implementation of CIA.

6.5.6 Using knowledge in real-life situations
CIA should expose learners to the application in real-life situations of knowledge and skills acquired in CIE and training. This can be done by linking CIE/assessment with workers' (intended) working activities. Learners may be requested to construct computer-integrated portfolios.

6.5.7 A paradigm shift

Most facilitators in the South African corporate training environment have noted the paradigm shift in assessment. Le Roux (2004) argues that the current assessment approach differs from the traditional approach in which learners were taught content and then subjected to an exercise at the end of a chapter or book in order to test their knowledge. This kind of assessment was not characterised by continuous assessment or assessment beforehand.

Le Roux (2004) refers to assessors who negatively pass judgement on their learners as 'butchers' because they aim at 'cutting learners to size'. 'Builders', on the other hand, are those assessors who cherish learners and support and develop them through OBET-aligned assessment.

Table 6.2 depicts a number of elements in these approaches to assessment (Le Roux 2004).

TABLE 6.2 BUTCHERS MODEL VS. BUILDERS MODEL

	Traditional practices of assessment 'Butchers model'		Alternative practices of assessment 'Builders model'
1.	Testing practices	1.	Constructive assessment
2.	Mainly summative in nature (focus is on final examination within a limited time)	2.	Formative (learning from assessment), summative (determining competence) and diagnostic (identifying strengths, weaknesses and prior knowledge)
3.	Educator is the sole judge of successful acquisition of knowledge (trainer-centred)	3.	Assessing learners' level of understanding within a content area (learner-centred)
4.	Surface approach to learning (meeting requirements of externally imposed tasks)	4.	Deep approach to learning (active search for meaning)
5.	Add-on (assessment not part of planning and design of learning programs)	5.	Integrated (assessment designed to match learning outcomes)
6.	Learning for assessment	6.	Learning from assessment
7.	Assesses learning content	7.	Assesses curriculum outcomes
8.	Tool for selection	8.	Tool for learning
9.	Caters for homogeneous student body	9.	Caters for diverse student body
10.	Limited range of assessment methods (one method is required to assess different kinds of competencies)	10.	Wide range of assessment methods to assess competencies, outcomes, processes and products

11.	Limited accountability to community of learners	11.	Optimum accountability to community of learners
12.	Limited credibility of results	12.	Greater credibility of results due to assessment by standards
13.	Incompatible formative and summative assessment techniques	13.	Consistency between formative and summative assessment

6.5.8 Critical and specific outcomes

The basis of CIA should be the specific and critical outcomes. This will enable the learner to progress towards the achievement of these outcomes at higher levels. Assessors should be guided by the range statements and assessment criteria (GDE 1999).

6.5.9 Learning activities and criteria

The performance indicators should be an integral part of CIA. They are stepping-stones to attainment of the critical and specific outcomes at higher levels. Assessors should choose criteria for each selected outcome to be assessed in the learning activity.

6.5.10 Valid outcomes

Most CIA assessors experience problems in identifying appropriate outcomes to be assessed in the learning activity (specific outcomes) and assessment criteria. GDE (2000) declares that identifying these outcomes and activities is difficult at first, but becomes easier when the assessor gains experience. CIA assessors should guard against having a set of fragmented and meaningless outcomes. Once activities have been designed, assessors should develop activities to assess skills, knowledge, values and attitudes. These tasks must be connected to the critical and specific outcomes. A lack of connection implies that CIA is not assessing the outcomes that assessors set out to assess. However, a valid CIA will assess what assessors intended to assess.

6.5.11 Learners with disabilities

GDE (1999) claims that every learner has some special learning needs. The Employment Equity Act (1998) requires that employers use the services of people with disabilities. Disability may sometimes be a learning barrier. Despite that, it is against the spirit of the Constitution to exclude learners with disabilities from learning opportunities. When CIE/assessment is designed and implemented, therefore, the needs of learners with disabilities should be catered for. In fact, the presence of these learners may be found through the analysis that is conducted prior to the design of the computer-integrated learning program. On the other hand, facilitators, assessors, subject experts, learning designers and other role players should always bear in mind that learners do not all necessarily learn in the same way and at the same pace. Computer-integrated learning programs and assessment should be characterised by a certain degree of flexibility.

6.6 ROLE-PLAYERS IN COMPUTER-INTEGRATED ASSESSMENT

People involved in CIA include facilitators, assessors, human resources development managers, line managers, learning designers, subject experts, project managers, development team leaders, critical reviewers, web designers and required SAQA role players.

The assessor and the learning designer have the main responsibility for the CIA. In most instances, the facilitator also acts as an assessor. The nature of the CIA will influence the composition of role players. Assessors/facilitators assess learners to support their progression and competence. This means that CIA requires a partnership between the facilitators, assessors, human resources development managers, line managers, learning designers, subject experts, project managers, development team leaders, critical reviewers, web designers and required SAQA role players. The direct integration will be between the learner and the assessor/facilitator and/or CIA program.

The learning designers and subject experts will mainly be involved during the design and evaluation phases of the CIA program rather than the implementation. In most organisations, line managers are regarded as promoters of their learning subordinates and thus follow the learner's progress. The consequence is high completion and success rates.

While there are several role players in CIA, it is of the utmost important to ensure the learners' rights to dignity and confidentiality.

6.6.1 SAQA role players

Traditionally, CIA would be confined to the assessor/facilitator and learner. However, SAQA (1995) specified role players that include the assessment advisor, assessor, moderator and verifier.

a) **Assessment advisor**

The assessment advisor advises employers on workplace issues around assessment. He/she enables workers to comprehend the rationale and processes of assessment. Workers need to know how they and the organisation benefit from CIA. In the corporate training environment, the need for an assessment advisor seems to have been exacerbated because most workers were not even aware that they had gained valuable skills, knowledge, values and attitudes over many years of employment. It is on this premise that these skills and this knowledge can be accredited towards a formal qualification. Of course, this is a separate process from CIA. There are about eleven official languages in South Africa, so it is essential for the advisor to use the language that learners will understand and communicate in.

b) **Assessor**

In most instances the computer-integrated facilitator will automatically be the computer-integrated assessor. The assessor should be well acquainted with the principles of good learning design, learning principles and lifelong learning. The assessor should work collaboratively with other role players in identifying opportunities for collecting evidence on learners' competencies. The assessment intentions should be communicated to learners. The assessor will keep assessment records and provide feedback and support to learners. He/she should communicate with other role players regularly. In the corporate training environment, the assessor should adhere to the Seta processes of registering learners and assessment results.

c) Moderator

The role of the moderator is to ensure that the assessment process is fair. The moderator is the custodian of the assessment quality management system within the organisation. Outside the organisation, quality assurance is carried out by the education and training quality assurer (ETQA). Just the same, it is the responsibility of the moderator to coordinate the external moderation and/or verification.

d) Verifier

The role of the verifier in the assessment process is the answer to the question of 'Who guards the guards?' The ETQA conducts the verification as part of the quality assurance within a particular industrial sector. One professional may play the roles of the advisor, assessor, moderator and verifier in different contexts.

6.6.2 Self-assessment

CIA allows learners to assess themselves. It is therefore important for the designers of CIE to expose learners to programs that empower them to assess themselves. The assessment program should allow the learner to reflect on his/her own performance. This kind of reflection will enable learners to improve and develop their competencies. The ability of learners to assess themselves gives them a certain level of independence. Adult learners should be able to set their own upper limits, particularly for specific and critical outcomes. They should also be able to produce their evidence of competence by putting together a portfolio.

6.6.3 Peer assessment

Learners should sometimes be involved in CIA by assessing each other's work. This is possible through computer-integrated discussion forums and chat rooms. It is advisable to empower learners with the necessary skills before they are allowed to engage in peer assessment. Computer-integrated peer assessment does not necessarily have to be formal. Negative judgmental statements and attitudes should be discouraged. Learners should be taught to be supportive and offer constructive critiques where and when necessary.

6.6.4 Organisational assessment

In a working environment, some colleagues may be requested to assist in CIA. For example, computer-integrated simulations, games, exhibitions and debates that have been developed by learners can be subjected to organisational assessment.

Once more, the learner's right to privacy and dignity should be considered before this is done. Assessors should empower the participants to coach and give constructive feedback to the learners. This kind of involvement may attract more workers to CIE and so lead to a broader appreciation within the organisation of CIA and the new learning/assessment paradigm.

6.7 TYPES OF COMPUTER-INTEGRATED ASSESSMENT

Types of CIA include formative assessment, summative assessment, baseline assessment, diagnostic assessment and system evaluation.

6.7.1 Formative assessment

Computer-integrated formative assessment entails a development methodology. The purpose of formative assessment is to monitor and provide support for learning progress. It is integrated into the learning activities continuously, requiring learner involvement. Well-designed formative assessment provides consultative feedback and guidance throughout the learning program.

The facilitator may organise follow-up learning activities after evaluating the computer-integrated formative assessment in a particular learning program. The nature of questions to which learners are supposed to respond should relate to particular learning outcomes and/or assessment criteria. Formative assessment will reveal aspects that may affect learners' competence. This enables the facilitator to determine the relevant developmental strategies, assessment tools and methods that are appropriate for the learning program and learners (Engelbrecht et al. 2000).

6.7.2 Summative assessment

Computer-integrated summative assessment involves a cycle of CIA activities which are derived largely from the performance record of the learner. It reveals how a learner has achieved a particular learning outcome. Summative assessment is also used to provide formative feedback to the facilitators (Engelbrecht et al. 2000).

6.7.3 Baseline assessment

The computer-integrated baseline assessment is employed by the facilitator at the start of a new learning program to determine learners' pre-knowledge, skills, values and attitudes regarding the learning area.

6.7.4 Diagnostic assessment

The purpose of computer-integrated diagnostic assessment is to uncover the kind and cause of learning difficulties. This is followed by relevant remedial assistance and guidance. Learning difficulties can be attributed to various factors. Learners who are more talented cognitively may require challenging learning activities, while the slow learner may be overwhelmed by this kind of work. On the other hand, learning materials may be badly designed, causing learners to find them difficult.

In a country such as South Africa, learners are confronted with socio-economic issues such HIV, poverty, unemployment and the legacy of apartheid. These issues can have a negative impact on the learning process. Because of this, some employers provide an employment assistance programme (EAP). The EAP is usually handled by social workers and/or psychologists. The CIA will indicate when specialist advice and support is required.

6.7.5   Systematic assessment

Systematic assessment is employed to evaluate the appropriateness of the organisational education system. This can be done by monitoring learners' achievement at regular intervals, using the organisation's goals and the country's legislation as measuring instruments. This information can be used for curriculum design and development. Education, training and development practitioners, role players and stakeholders may have to reconsider their goals, needs, strategies and tools. In some organisations CIE was introduced after the systematic assessment.

CIA must be continuous. The rationale is to support and provide feedback to learners and this should be integrated into CIE.

6.7.6   Integrated assessment of applied competence

According to the University of South Africa (Unisa 2004), integrated assessment has the following qualities:
• Assessing a number of outcomes together
• Assessing a number of modules together
• Using a combination of assessment methods and instruments for an outcome or outcomes
• Collecting naturally occurring evidence such as in a workplace setting
• Acquiring evidence from other sources such as supervisors' reports, testimonials, portfolios of work, logbooks, journals, etc.

The Unisa report (2004) adds that applied competence refers to the foundation and the practical and reflective elements of assessment. This implies that learners should be able to demonstrate the knowledge, skills, values and attitudes that they have acquired in a particular computer-integrated discipline or field of study (foundational knowledge). Learners should also be able to apply these competencies in a given situation. Integrative assessment includes extended and complex open-ended learning activities that require the integration of the assessment criteria and specific outcomes (GDE 1999). Aspects of selected specific and critical outcomes should therefore be addressed through integrative assessment. This approach should be an integral part of the CIA design and its implementation.

6.8 RECORDING COMPUTER-INTEGRATED ASSESSMENT

The success of CIA depends on sound and thorough methods of recording learner performance over a period of time. Computer-based observation sheets can be used to store learner's progress towards achieving a particular learning outcome. The observation sheets may also be print-based (off-line). The data reflected in the observation sheets is made up of assessor's observations and learner's involvement and performance in CIA.

The assessor must ensure that a CIA program is based on a particular assessment criterion. The CIA should be:
- uncomplicated and easily interpreted by the assessor and other stakeholders
- flexible enough to accommodate the addition and deletion of information when the need arises
- a genuine, factual indication of learners' strengths and areas of support needed
- comprehensive enough to demonstrate learner progress
- ongoing and continuous
- helpful in the reporting process
- readily accessible
- kept in a secure place to protect the confidentiality of learner progression (GDE 1999)

The assessment criteria and performance indication must be reflected in a particular session of CIA. The assessment records should be manageable and easier to access. On the other hand, records should be used by assessors to provide learners with feedback and support. The feedback should describe learners' strengths and weaknesses. Feedback should not be accompanied by detailed descriptions of learner performance.

6.9 REPORTING LEARNERS' PERFORMANCE

CIA cannot be conducted for its own sake. It should therefore be characterised by regular reporting. The reports should provide learners, managers, facilitators and other role players with significant information that can be used for various teaching and learning purposes (GDE 2000) including:
- describing and detailing the learning that has taken place and the complexity of the learning achieved
- outlining for subsequent facilitators the learners' strengths and support needed, which ensures the continuity of computer-integrated learning programs
- enabling managers to participate in the learning process of their subordinates
- making facilitators more accountable to learners, the organisational educational system and the wider community

Employees and their managers should be part of the CIA process. Facilitators should manage access to CIA records and reports. Reporting can be done in various ways (GDE 2000). These include verbal presentation, written summation and short notes. Reporting may be formal or informal, impromptu or planned, and presented in general or specific terms. The learner's performance should not be confined to marks, percentages and symbols.

6.10 SELECTING A BLENDED ASSESSMENT METHOD

The assessor may blend assessment methods. Blended assessment consists of a combination of methods. These include observation, written, formal examination, journal, logbook, project, portfolio of evidence, self-assessment, and peer assessment. (The details below have been adapted from CETA 2001).

6.10.1 Observation assessment method

The observation assessment method consists of on-job demonstrations of applied competence. The assessor may observe the actual performance and product directly or use video-recording. In this instance, performance is assessed in situations that simulate real conditions as closely as possible. This kind of assessment provides authentic and valid evidence because performance is carried out in the work situation. It also enables continuous assessment. Learners find observation assessment methods meaningful since they may form part of routine workplace supervision. They are cost effective and minimise disruption to the workplace. Despite that, the assessor may need an extra source of evidence to assess the breadth or variety of performance in different contexts, as per the range statement in a SAQA-registered standard/qualification. In some instances, this method is artificial and may not fully reflect workplace conditions or pressures.

6.10.2 Written assessment method

In the written assessment method the learner is expected to provide a detailed description to solve a problem in the scenario or to give a plan of action for it, for example when an emergency situation in the workplace is described. Understanding of evacuation procedures, etc, is assessed, as well as the candidate's ability to review procedures. This method provides evidence of learner's knowledge of procedures, as well as his/her problem-solving skills. It is useful in assessing reflexive competence and integration of knowledge into performance when access to a real situation is impractical. However, it is not real and learners may react differently in real circumstances. The learner's response is also passive, in contrast with role play, where the learner acts out the situation. It is therefore important for the assessor to describe the scenario in a very lifelike manner and his/her record should be written carefully to reflect the assessment criteria.

6.10.3 Formal examination

Formal examination is a method that is in line with what people expect of traditional assessment. It is characterised by a set of oral or written questions about an area of learning or performance, often set by an external assessor. This requires extended written answers or oral explanations. Formal examination is usually undertaken in standardised conditions with a time limit and no previous knowledge of questions. This type of assessment allows the standardisation of results across large groups. It is also useful for providing evidence of scope of knowledge. The limitation of formal examination is that the emphasis is put on exam skills, and not applied competence. It discriminates in favour of people with good communications skills. The assessment results should consequently be moderated. The assessor will probably need more evidence to assess applied competence.

6.10.4 Journal assessment method

Journals can be used with logbooks or a training register. It is a self-reporting method to generate evidence of performance, progress, experiences, attitudes and personal feelings in an ongoing situation. It is a useful way of assessing progress, achievement or performance. Journals and logbooks have clear specifications and give guidance on how essential information is to be recorded. This method is a good way to reflect practical, foundational and reflexive competence. It develops good record-keeping discipline/habits in the learner and encourages self-motivation. At the same time, it takes time to compile and assess. It may be difficult to ensure that the logbook is a true record of achievement and experience. Logbook entries may be faked and their reliability is hence questionable. The assessor would have to review the logbook with the learner to ensure satisfactory performance.

6.10.5 Project, assignment and report

This type of assessment is characterised by a single task or set of tasks that combine a range of competencies and that result in a written or oral report, assignment or presentation, or a mode, product or portfolio of evidence (or combination of these). It may involve research and collection, analysis, reporting on and application of information. The project assessment method can provide evidence of foundational and reflexive competence, and practical competence in some instances. It allows a learner a lot of choice in preparing the evidence to his/her own taste and preferences. This kind of assessment may require high-level research, organisation and communication skills. It may be difficult to find out whether it is the learner's own work. Good presentation or packaging may attract the assessor's attention more than learner's performance and content of the project, assignment or report. The assessment criteria should be the main guide in measuring outcomes. Not only should the final product be assessed, but also the process that the learner followed to reach the end product.

6.10.6 Portfolio of evidence

In this assessment method, a collection of evidence is presented in different ways in one portfolio. The portfolio is supposed to contain confirmation of personal, performance, skill or life experience. Documents that record previous experience, for example certificates, testimonials, or letters of recommendations, may form part of the portfolio. The dilemma of the portfolio of evidence is that it may be difficult to find out whether it is the learner's own work. Then again, a good or bad style of presentation may influence the assessor's opinion. The assessor needs to consider whether the evidence is outdated or current. Relevant assessment criteria should be clear and kept in mind. Assessment of the portfolio may require further explanation or clarification of evidence through an interview or other forms of assessment.

6.10.7 Self-assessment method

This technique is used so that a learner can find out his/her own level of competence. It makes it possible for the assessor to find out what the learner thinks his/her level of competence is, and shows where there are gaps between what it should be and what it is. It also enables the assessor to determine the learner's readiness for assessment. Learners have an opportunity to look at all the assessment criteria in relation to their levels of competence. Self-assessment supports the assessment decision and provides the basis for discussing areas of perceived competence. Areas where the learner thinks he/she is competent, but is not yet proficient, will be identified and addressed. Yet, it may not be an accurate assessment because some learners underestimate their level of competence, while others have high opinions of their ability. Assessment criteria should hence be clear for valid self-assessment to be made.

6.10.8 Peer assessment method

This assessment method involves obtaining a second opinion from co-learners on applied competence. It provides extra evidence to support direct evidence collected through another assessment method. It is very useful in assessing group competence where 'peers' are in an ideal position to assess skills, such as workers assessing managers. Peer assessment method enables the assessor to look critically at assessment when views of peers differ in certain areas. Still, it may not be accurate because of bias or favouritism by peers who are not willing to give an honest assessment. It may also be threatening to the candidate and puts peers in an unenviable position if the competence observed is not achieved when they work with the learner.

6.11 PRINCIPLES OF STANDARDS-BASED ASSESSMENT

To ensure good quality within the SAQA national standards-based assessment processes, certain principles need to be used in all design and implementation of CIA. Assessment should:
- be clearly linked to learning and teaching
- be designed to meet the clearly stated purpose
- be clear and understandable to learners
- be derived from registered national standards and qualifications
- comprise a contextually relevant mix of criterion, norm and ipsative approaches
- use a purpose-driven optional mix of assessment tools, both summative and formative
- credential the outcomes of integrated applied competence
- be reported in a rich and relevant way
- clearly demonstrate its contribution to transformation and equity
- be credentialed only with the NQF-registered credits at the appropriate level from specific standards and qualifications
- have fully developed policies and procedures including provision for appeal

- be carried out only by appropriately trained and ETQA-registered specialists at all levels
- be moderated from time to time and amended to reflect the changing needs and state of good assessment practice
- be recorded and accessed in such a way that the credits do not reflect how, where and when they were acquired (CETA 2001)

6.12 COMPUTER-INTEGRATED ASSESSMENT AND THE ELIMINATION OF BIAS

Several factors can lead to bias in assessment (GDE 1999), including the digital divide, language, gender, culture and other discriminatory issues. As common barriers in making assessment decisions, CETA (2001) identifies the halo effect, first impressions, the contrast effect, stereotyping, a similar-to-me attitude, and giving more weight to positives than to negatives.

6.12.1 The digital divide

The South African corporate experience has revealed that it is pointless to impose CIA on learners who do not have access to computers. These learners would be disadvantaged by their lack of accessibility to computers (the digital divide). It is therefore important for any organisation to ensure that all learners who are supposed to participate in CIA have some access to (networked) computers. This can be done by cooperating with corporate, government and non-governmental organisations (NGOs) which provide computers to communities and schools. Community centres and computer laboratories may be made available for adult learners in the evenings and at weekends to allow them to participate in CIE activities.

Private and public partnerships that are providing computers to schools and community include the GautengOnline project, Khanya Technology in Education, the Telkom 1000 schools project, ThinkQuest, the Marconi-Department of Education ICT project, SchoolNet SA, Digital Partnership South Africa, and Africare. These computer resources/projects should be opened to the whole community and learners who work in the corporate environment, so as to minimise the digital bridge. Stakeholders such as the government, NGOs, corporates, Telkom and Eskom should work together in endeavours to bridge the digital divide.

6.12.2 Language issues

In the corporate learning environment, CIA designers and providers assume that every learner understands and communicates in English. This assumption and practice is very flawed and discriminatory. Organisations should endeavour to provide education, training and development in the official languages of learners. Workers in management prefer to learn and communicate in English rather than their home languages. Programs in some African languages will therefore become redundant. Some human resources development managers argue that designing learning programs in all eleven official languages is an expensive, albeit politically correct action that does not necessarily add value to teaching and learning purposes. The issue of language is also determined by the profile of targeted learners, so learner analysis before the actual design is very important.

With the exception of English and Afrikaans, the current official South African languages were neglected during the apartheid era (Madiope and Dagada 2004). Nevertheless, where learners are being taught to communicate in a particular language via CIE and assessment, the medium used would be that language. There is political and moral pressure on South African organisations to develop and use the previously neglected official languages for communication and learning. Most organisations have therefore established language policies and guidelines.

Language issues may cause many frustrations in CIA. This could be exacerbated because other languages are spoken in South Africa besides the official languages. Sentence construction in South African languages is determined by cultural issues. This has been demonstrated by the African Management Programme (2003:53):

The cultural mind is reflected not only by words but also in sentence structure. There is a fine South African example to illustrate this: in the Western cultures there is always someone to blame when something goes wrong. This is not so in the black cultures. Something can go wrong or get lost without someone necessarily being responsible. This is reflected in the sentence structure. For example: A person from the Western culture would say: 'I have lost my knife'. In Sesotho one would say: 'I have been lost by my knife'. The person thus has no share in the behaviour of his knife.

Assessors and learners who are participating in CIA should avoid expressions commonly used in their home language that may not be understandable to other role players. The African Management Programme (2003:55) refers to this as jargon. People who use jargon assume that other people will comprehend it, whereas this is not usually so.

6.12.3 Gender issues

CIA should address learners objectively without favouring a particular gender. In South Africa, an assessment program that is biased towards a particular gender will be viewed as inconsistent with the Constitution (1996). GDE (1999) claims that it is difficult not to have certain prejudices and expectations about a particular gender. The stereotypes society has exposed us to, consciously or unconsciously, cause us to respond differently to a person of a certain gender. It is therefore incumbent on assessors to eliminate this in computer-integrated practices.

On the other hand, discriminatory practices in the workplace – such as using largely female workers to do secretarial jobs, make tea and clean – will have psychological effects on their performance in CIA. An organisation should therefore have a gender policy that will have a positive impact on education, training and development activities.

6.12.4 Cultural issues

The learning developers of computer-integrated learning programs and CIA should recognise and respect the diversity of their learners. This should be done by being sensitive to and tolerant of the culture, race, ethnicity, nationality and beliefs of other people in the computer-integrated learning environment. For example, learning developers and assessors should shy away from providing examples, scenarios and tasks that promote only one religion. Such biased presentations cause some learners to be advantaged and others to feel rejected. The Constitution makes provision for diverse cultural, religious or linguistic uses and practices.

In a blended learning environment these issues can be very thorny and difficult to handle. The extract by the African Management Programme (2003:53) below provides an example of complications and challenges related to diversity. The computer-integrated and blended learning environments are not immune to these, which should be attributed to translation problems.

a) Translation problems in marketing communication
  - An American airline operating in Brazil advertised plush 'rendez-vous lounges' on its jets, only to discover that rendezvous in Portuguese means a room hired for love-making.
  - General Motors' 'body by Fisher' was translated as 'corpse by Fisher' in Flemish.
  - Colgate's 'Cue' toothpaste had problems in France as cue is a crude term for 'butt' in French.
  - In Germany, Pepsi's advertisement, 'Come alive with Pepsi', was translated as 'Be resurrected with Pepsi.'
  - Sunbeam attempted to enter the German market with a mist-producing curling iron named the Mist-Stick. Unfortunately, mist was translated as 'dung' or 'manure' in Germany.
  - Pet milk encounters difficulties in French-speaking countries where pet means, among other things, 'to break wind'.
  - Fresca is a slang word for 'lesbian' in Mexico.
  - Esso found that its name phonetically meant 'stalled car' in Japanese.
  - Kellogg's Bran Buds translates to 'burned farmer' in Swedish.

The complications associated with cultural differences are reflected in another extract by the African Management Programme (2003:33).

b) Gestures
  In Atlanta waiters were taught the real meaning of gestures for different nationalities that converge there for the Olympics in July 1996.
  - A circle made by the thumb and the first finger is usually meant to be 'It's OK', but for the Germans it means 'Number 1', for the Japanese it means 'Five', and for the Arabians 'I won'. To the Chinese and Australians it has a rude or negative meaning.

- An open hand held up means 'wait awhile' to most people. To use this for a West African is not advisable as it means that he has five fathers! And for Greeks it has a meaning that is only a little less rude.
- Rubbing the thumb and first finger is taken to mean money, but the French use it to mean 'perfect' – while to some Europeans it is an extremely rude and vulgar gesture.
- Two fingers held up to indicate 'two open seats' must not be used at all!
- Waiters are cautioned that smiling Japanese people may be very angry.

Gestures have meanings that can cause communication problems, especially in a blended learning environment where participants are able to observe one another's body movements (especially hands and arms).

6.12.5 Other anti-bias issues

GDE (1999) insists that facilitators and assessors should accept that learners who are perceived differently owing to their 'physical, psychological, emotional, socio-economic, cultural or other characteristics' should be provided support generally and specifically in CIA. If this is not done, these learners will be disadvantaged, and the integrity of that particular assessment will be questionable and unfair.

6.12.6 Human rights

The Constitution (1996) seeks to promote the equality and dignity of each person. This imperative should be considered when CIA is designed and implemented. Legally, assessors who ignore the Constitution can be prosecuted. GDE (1999) demands that no learner should perform poorly as a consequence of 'intimidation, rejection, or low expectations' on the part of the assessor and/or facilitator.

6.12.7 Common barriers

CETA (2001) identifies barriers of computer-integrated decisions such as the halo effect, first impressions, the contrast effect, stereotyping, similar to me, giving more weight to positives than negatives, the experimenter effect, and grading.

a) The halo effect

The computer-integrated assessor should make decisions about learner performance based on current behaviour rather than previous performance.

b) First impressions

The assessor should pay attention to all stages of assessment rather than form a premature impression about the learner, based on the first stages of CIA.

c) Stereotyping

The assessor should make decisions about learner performance based on the performance, rather than other features of the learner.

d) Similar to me

The assessor should make fair decisions, regardless of whether a learner has characteristics that are similar to or different from him/her.

e) Putting more weight on positives than negatives

If the learner performs exceptionally well in a particular aspect of the CIA, the assessor may tend to have unrealistic expectations of the learner in all aspects of assessment.

f) Experimental effect

The presence of the assessor in a blended learning environment can sometimes have a negative impact on the outcome of the assessment. This is more problematic when learners behave differently when their assessment tasks are being observed.

g) Grading

Grading CIA in a standard-based system is a highly contentious issue. Some assessment experts claim that 'assessment is assessment', and no other interpretation should be made. However, others claim that some learners are more able than others, and this should be recognised. Some human resources development specialists in the South African corporate training environment have decided that computer-integrated assessed competence in the workplace should possibly not be graded because workplace assessment shows the achievement of competence. Competence should be declared only once the assessor is convinced that the learner being assessed is definitely competent.

6.13 LEARNER-PACED ASSESSMENT AND PROGRESSION

The assessor should practise learner-paced assessment and progression by giving credit for outcomes achieved, tracking progress to identify needs, providing different learning activities to respond to different needs, resisting labelling and streaming, and monitoring progression of learners (GDE 1999).

6.13.1 Credit for outcomes

Traditionally, learners have been given credits according to the time that they have spent in learning institutions. Experience in the corporate training environment has revealed that it makes better educational sense to award credits or qualifications for the outcomes that the learner has achieved rather than the period spent in a particular learning environment. This means that once learners have satisfied the requirements for an outcome at a certain level, they can then progress to a higher level. However, if they cannot apply the acquired knowledge, skills, values and attitudes, they should return to the previous outcome before they can move forward. It is therefore important for assessors and facilitators to recognise that learners who are participating in the same computer-integrated learning program are not necessarily at the same level. In some instances they should be given different learning and assessment to accommodate different learning needs.

6.13.2 Tracking progress to identify needs

Because learners in the same computer-integrated learning program may be at different levels in terms of abilities and learn differently, OBET is learner-centred. In computer-integrated OBET, learners can learn at their own pace. Assessors should therefore ensure that CIA is learner-paced and the progress of learners is constantly tracked. This will allow the assessor to identify learner's needs. The GDE (1999) suggests when assessing a learner against a certain outcome an assessor should ask: 'At what level is the learner performing?' It is not enough to ask, 'Can the learner do this?', 'Does the learner know this? or 'How well and at what level can the learner apply this knowledge?' This will reveal the needs of learners that should be addressed, and assessors will be able to record the information.

6.13.3 Different activities in response to different needs

Although facilitators and assessors should try to serve different needs of learners, it is difficult and costly to design tailor-made computer-integrated learning programs for each learner. The GDE (1999), however, suggests that assessors who provide outcomes-based CIA 'can allocate follow-up tasks and activities on the basis of (for example) three broad groups: those who have finished a given activity successfully and quickly; those who are still engaged in it; and those who cannot cope with it'. The first group may move forward fast by tackling increasingly demanding tasks, showing their capability against a particular outcome at a higher level than is expected of their normal grade. The second group may be given tasks that are of a lower level than the first group, while the third group will be engaged in activities that are lower than those of the second group. A lot of time may be required to enable the second and third groups to perform like the first group. This will also require that the two last groups engage in serious practice.

6.13.4 Resisting labelling and streaming

Groups should not be permanent. The danger of creating 'streams' or 'bands' is that it may lead to generalised low or high expectations by assessors and other role players and therefore negative psychological implications. A gifted learner who has been mistakenly branded a 'slow learner' may lose confidence and began to perform poorly. This can be prevented by starting every new set of learning activities with a common computer-integrated baseline assessment, so that the assessor can have a broad picture of the learners' needs in a particular program.

6.13.5 Monitoring progression of learners

In an ideal situation, learners who are participating in the same computer-integrated learning program should proceed to the next learning program together. However, this is not possible if particular learners fail to meet the requirements of a specific learning outcome. Some learners do not complete the learning programs owing to unforeseen circumstances. The practice in some organisations is that if a learner has to repeat a learning program, documentary evidence from the facilitator, assessor and line manager is required. This should reflect in detail the learning problems that the learner has experienced. Possible solutions to these problems should be suggested.

6.14 ARGUMENTS FOR AND AGAINST COMPUTER-INTEGRATED ASSESSMENT

Several claims have been made for and against CIA (McLean 2002:7). These claims will be gauged against the experience of the corporate training environment regarding CIA.

6.14.1 For computer-integrated assessment

a) High level of reliability

CIA offers a high level of reliability in marking assessment activities. This claim was found to be true of the South African corporate training environment, especially in multiple-choice and short questions.

But CIA programs cannot deal with questions that ask learners to define, describe and discuss issues in their own words. On the other hand, when learners want to justify their responses to questions that need 'yes' or 'no' answers, the program does not provide such opportunities. Assessors found this problematic since it could lead learners to answer questions by merely guessing. Even so, the perceived shortcomings of CIA should not eclipse its ability to mark assessment activities objectively.

Some of the administrative work related to assessment can be handled by CIA. This includes calculations and record keeping. CIA can identify and inform the assessor of the performance of a particular learner, for example by flagging names of learners who have learning difficulties.

The ability of CIA programs to mark assessment activities is enhanced because a computer does not become tired, bored or moody. This gives CIA a lot of clout when it comes to integrity. Although the human assessor may have a lot of integrity, he/she may make mistakes, and thus CIA is more reliable.

b) Flexibility

Traditionally, assessment can be conducted only on specified dates and at specified times. In other words, those who are involved in assessment have to adhere strictly to the test, assignment, homework and examination schedule. Once the schedule has been approved by a relevant authority, it becomes final and cast in stone. However, in some organisations there is a move towards flexible assessment. In fact, flexibility is one of the reasons for employing CIA. Assessors do not see the need to wait for a particular date to assess learners if they are ready.

CIA flexibility is not restricted to time, but also allows those who are involved in assessment to participate wherever they are. In other words, learners and assessors may not necessarily be at the same venue. Learners need not always go through the whole CIA program, but may do a specific section within the program.

c) Cost-effective

The ability of computer-integrated assessment to mark assessment activities and keep records automatically saves a lot of money for organisations. Paper-based assessment, on the other hand, demands a lot of money and time. Where thousands of workers do paper-based courses, a lot of money is spent on paying markers. Other payments are made to assistants who do miscellaneous work and record keeping. Paper-based assessment, as opposed to CIA, is characterised by logistics that require huge funding. These may include transporting question papers and scripts from one point to another and storing question papers and marked scripts. Some organisations require marked scripts to be kept in storage for at least three years. This can be very expensive.

In some instances, scripts have to be marked at a central venue that has to be paid for. The organisation may also have to pay the accommodation costs of those involved in assessment, especially the assessors, markers, moderators and assistants. There are also costs associated with the production/purchase of question papers and answer books. CIA is thus perceived to be cost-effective compared with paper-based assessment. Of course, the value and quality of learning eclipse costs.

There are several reasons other than cost for employing CIA or paper-based assessment.

d) Immediate and detailed feedback

Computer-based assessment enables learners to obtain immediate feedback, which enables learners to maintain momentum. During interviews, assessors and learners complained of the agony of waiting for feedback in paper-based assessment. Waiting for feedback for a long period can be traumatising, especially if passing the assessment is linked to possible promotion. CIA is highly regarded for providing feedback speedily. Some programs provide very detailed feedback, which enables those involved in assessment to conduct quantitative and qualitative analysis. After this analysis, learners who are experiencing learning difficulties can be assisted. Assessors may also effect changes in the program itself because of the analysis. Thus the analysis may have a ripple effect in a particular computer-integrated learning program.

e) Multimedia elements

CIE has the ability to employ multimedia elements. The South African corporate training environment revealed that multimedia elements (sound, colourful graphics, moving images, video and audio technologies) are relevant in a CIA program linked to work-integrated learning. This view is supported by McLean (2002:8). Soldiers, for example, who are supposed to be assessed on using explosives, may demonstrate that ability through computer-integrated simulations. It is easier to do this kind of demonstration through computer-integrated assessment than paper-based assessment.

It is impossible to use the multimedia elements in paper-based assessment. CIA enables assessors to use trial-and-error, games and simulations. This enables learners to apply what they have learnt and are being assessed on in a real-life work situation. The use of multimedia elements can reduce anxiety among learners. Learners find the assessment enjoyable, so their concentration and interest are retained.

6.14.2 Against computer-integrated assessment

a) Assessment of higher-order thinking skills

Although CIA is very efficient in assessing through 'objective tests' such as multiple-choice assessment, matched-item assessment, drop-and-drag labels and other objective tests that require short answers, it has been posited that it cannot adequately assess higher-order thinking skills. After conducting interviews and observations in the corporate training environment, one finds this claim very debatable. Objective questions can be used to assess higher-order thinking skills. But it is impossible to use CIA in questions that require learners to write essays, describe, define, critique and discuss, because marking these types of answers requires considerable subjectivity on the part of the assessor.

b) Costs

The inclusion of multimedia elements in CIA can be grossly expensive and even more so if video is used. However, CIA that does not include multimedia elements is not expensive to develop. Off-the-shelf programs are relatively cheap. McLean (2002:8) argues that the costs of assessment are not confined to its development but also include administration. CIA is cost-effective if provided by centralised providers nationally and internationally. The problem with off-the-shelf and centrally provided CIA programs is that they are not customised for a particular country or for organisational needs.

c) Accessibility to computers and infrastructure

CIA can be employed in a learning environment that has computers. In some instances these computers should be networked. This was found to be problematic in South Africa, since most people do not have access to computers. Vast areas of the country do not have electricity or a telecommunication infrastructure, so computers cannot be networked. During observations and interviews, it was found that bandwidth was a very serious problem. Bandwidth in South Africa does not have sufficient capacity. Also, where computer facilities and infrastructure were available, facilitators and learners did not always have the skills to use them.

d) Information security and cyber-crime

CIA is not immune to problems related to information security and cyber-crime. Verton (2002:16) and Schneier (2000:307) claim that information system crimes and cyber-attacks were expected to increase from 2003. However, the government and business have taken measures to protect electronic commerce and education activities. These measures are just not protective, but also developmental. The South African National Intelligence Agency (NIA) has offered to assist the business community to fight industrial espionage and attacks launched by foreign intelligence agents pretending to be private security companies. Mamaila and Green (2002:1) cite the example of a South African company that employed the services of a foreign security company, which stole strategic business information and sent it to its mother country in Europe.

The Electronic Communications and Transactions Act (ECT Act) was passed in 2002. While this act is perceived to be designed primarily to protect online merchants and consumers, it will also protect CIE, training and development activities. It contains specific definitions of the nature of the security an organisation should provide for its website.

6.15 HOW WILL COMPUTER-INTEGRATED ASSESSMENT BE GRADED?

When an assessor designs CIA, he/she should consider what will happen after learners have completed the assessment (Horton 2000). He/she should know how learners' performance will be evaluated and what kind of feedback will be provided. Table 6.3 provides some possibilities, accompanied by their advantages and disadvantages.

TABLE 6.3 POSSIBILITIES, ADVANTAGES AND DISADVANTAGES OF ASSESSMENT

Technique	Advantages	Disadvantages
Answers are evaluated by a script or program on the learner's computer	Evaluation is immediate No network connection is required. Works for courses on CD-ROM or hand disk The computer is non-judgemental. Learners do not fear human criticism	Limited to simple forms of evaluation Assessor cannot monitor learner's progress
Answers are transmitted to a remote computer, which analyses them and generates a response	Assessor can monitor learner's progress Evaluation is quick The computer is non-judgemental	Requires network connection Limited to simple forms of evaluation
Answers are e-mailed to the assessor, who grades them and writes back an evaluation	No limit to the kinds of questions Assessors can spot learners' subtle misconceptions Assessors will have knowledge necessary to evaluate learners	Quality of evaluation depends on the knowledge of the assessor Learners have to wait for a response Requires extra work by assessors Learners may feel reluctant to expose their ignorance to the critical gaze of the all-powerful assessor
Learners have a co-worker or on-site advisor examine their answers and comment on them	Co-workers can show how something applies to the learner's real-world activities	The co-worker may not be available. The learner may have to print out the screen Co-workers may lack knowledge and expertise necessary
Learners evaluate their own work using a procedure spelled out by the assessor	Having the learner find the answers in the preceding material provides a second learning opportunity	The answers are not easy to find Learners will consider searching for answers a waste of their time
Other learners evaluate the work	Peer evaluations help learners develop judgement skills. They can also foster a sense of teamwork	Learners may lack the necessary knowledge Some learners are not mature enough to politely and objectively evaluate their peers, with whom they feel they are competing

Source: Adapted from Horton (2000:276)

CHAPTER 7

COMPUTER-INTEGRATED ASSESSMENT: MICRO, PEDAGOGICAL AND TECHNICAL DESIGN

CONTENTS

7.1 INTRODUCTION

The theoretical foundation for this chapter is derived from Horton (2000), and the remaining details reflect the author's experience and data collected through observations and interviews in the South African corporate training environment.

This chapter outlines types of questions that can be employed in CIA. These include true/false, multiple-choice, text-input, matching-list, click-in-picture, drag-and-drop, simulation, and fill-in-the-blank questions. Each type of question is accompanied by a practical example. Some types of question require advanced technology, broad bandwidth, multimedia and lots of plug-ins.

7.2 TRUE/FALSE QUESTIONS

True/false questions require that learners choose whether a statement is either true or false. The questions below are examples of true/false questions from a subject on the provisioning of human resources.

THE PROVISIONING OF HUMAN RESOURCES

Answer the questions below by clicking whether a statement is true or false.

1. Human resources management deals with the management of people in an organisation.
   ☐ True   ☐ False

2. Job analysis may be described as another method of communication.
   ☐ True   ☐ False

3. A job analysis has two components; a job description and a job specification.
   ☐ True   ☐ False

4. A job description define the qualification, experience and personal qualities required by the job.
   ☐ True   ☐ False

5. Human resources forecasting deals with the estimation of the demand for employees with certain skills.
   ☐ True   ☐ False

(EXIT)   (HELP)   (MENU)   (BACK)   (NEXT)

7.2.1 Formulating true/false questions

True/false questions should be employed in CIA when the assessor wants to make definite judgements. They compel learners to make decisions about certain matters:

- Is a statement correct or wrong?
- Will a procedure function or not?
- Is a procedure safe or dangerous?
- Does a proposal comply with standards?
- Should you accept or reject a proposal?
- Which of the two options should I choose?

Before the assessor employs a true/false question, the kind of question should be considered as well. True/false questions tend to be confined to cases, so some learners may resort to guessing. During interviews and observations in the corporate training environment, it was noted that the manner in which these questions is formulated is very critical. Assessors in the corporate environment therefore try to formulate true/false questions in a way that encourages thinking rather than mere guessing.

7.2.2 How to formulate true/false questions

- The assessor should ask several questions on a topic. This will make it difficult for the learner to guess them all correctly.
- For each topic, the assessor should formulate true/false questions in various ways so that the correct response is sometimes 'false' and sometimes 'true'.
- The assessor should formulate the question in neutral terms so that the response is not implied.
- The assessor should give clear hints and/or explanations for incorrect responses.

The assessor should formulate the questions to fit the responses. Questions should not be complicated and difficult to comprehend. True/false questions should shy away from asking learners what they perceive. On the other hand, the assessor can discourage guessing in various ways:

Guessing can be discouraged by penalising learners for it. In awarding scores for true/false questions, the assessor should give 1 for correct answers, 0 points for unattempted questions, and minus 1 point for incorrect responses. This means that guessing could be worse than not responding.

The assessor should make it difficult for the learner to get higher scores by merely guessing. The possibility of getting 5 of 10 true/false questions correctly by guessing is 50%. On the other hand, getting 80% correct by plain guessing is just 20%.

The assessor could discourage guessing by posing more questions. If the assessor increases the number of true/false questions to 20, the possibility of obtaining 80% correctly by sheer guessing decreases to less than 1%.

7.3 MULTIPLE-CHOICE QUESTIONS

Interviews and observations in the South African corporate training environment revealed that multiple-choice questions in CIA allow the assessor to provide a list of possible answers for the learner to select from. While these questions are easy to formulate and to comprehend, they tend to tempt learners into guessing rather than thinking more deeply.

There are various kinds of multiple-choice questions. These include pick-one and pick-multiple.

7.3.1 Pick-one questions

Pick-one questions enable learners to respond by picking one answer from a list of alternatives. Only one response will be correct. In the example, learners are provided with steps in the employment selection process, and they have to pick the one combination of steps that is suitable in the selection process.

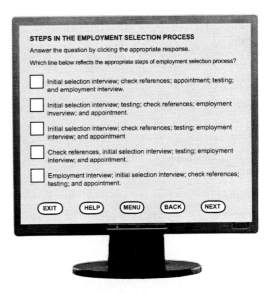

7.3.2 Pick-multiple questions

Pick-multiple questions allow learners to pick one or more responses from the list of alternatives. In the example, learners are given an opportunity to pick all the answers that are right.

Multiple-answer tests accommodate more advanced questions than are possible with one-answer or true/false tests. The learner is therefore required to make a series of related decisions.

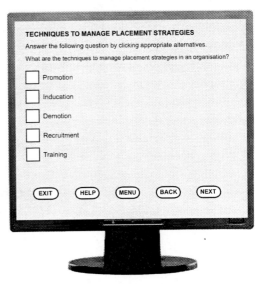

7.3.3 Alternative forms of multiple-choice questions

In CIA, the assessor presents multiple-choice question options using scrolling buttons. If the assessor wants to ask many multiple-choice questions about a particular subject or there is limited space, the scrolling buttons are more appropriate. The example on the subject of the maintenance of human resources, demonstrates how scrolling buttons can be employed.

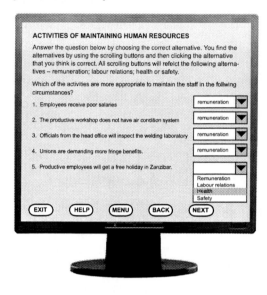

7.4 TEXT-INPUT QUESTIONS

The text-input question is a type of assessment in CIE that allows the learner to type in the response to a question. Usually this entails short responses to objective questions. In the example below, on the subject of labour relations management, learners are required to type in four types of Storey's role.

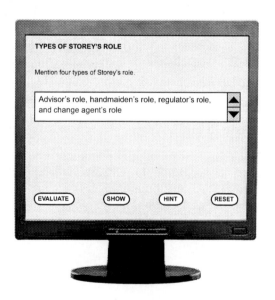

7.4.1   Formulating text-input questions

The most difficult part of formulating text-input questions is phrasing them in a manner that enables the CIA program to assess the answer. The assessor should therefore:

- Formulate the question in a way that limits the number of correct answers
- Formulate the question so that the answer can be assessed according to the availability or unavailability of specific words or phrases, but not necessarily the exact order or syntax of the response
- Formulate the question in a manner that accommodates synonyms, grammatical differences and common spelling mistakes
- Inform learners how to formulate their responses and, if necessary, provide an example of a well-formulated answer
- Make it clear in the instructions whether the answer should be in the form of text and/or numbers
- Divide a complex question into separate questions, enabling learners to provide a short specific answer to each (the assessor should not pose two separate questions to be responded to in one input space)
- Inform the learner of the format and other requirements via free-form input (if the format is not clearly stated, learners may provide answers that will not be marked by the CIA program)

Automatic scoring of free-form text is hard to implement practically. It is difficult for CIA programs to scan the learner's answer appropriately.

7.5 MATCHING-LIST QUESTIONS

Assessors in the corporate training environment allowed learners to indicate which items in one list match items in another. In the example on the subject of labour relations management, scroll-down lists let the learner select answers that match the list of alternatives. The learner can choose from the list on the left or the one on the right. These lists are synchronised.

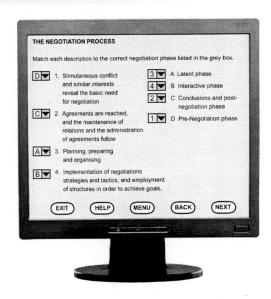

7.5.1   Formulating matching-list questions

The assessor should make matching lists simple to enable learners to concentrate on the relatedness of items in the two lists. The assessor should also:

- *Clear list items*: The assessor should use simple terms or a glossary that will allow the learner to look up terminology.
- *Keep the lists short*: Lists should be short so that they can all be accommodated in the same display. However, if they do not fit into the same display, the learner should be provided with buttons to move backwards and forwards.
- *Learners indicate matches easily*: Instead of learners typing the letter or number of the corresponding item, let them choose it from a list of alternatives, and drag and drop items on one another. Alternatively, the assessor can put lines between items.
- *Process of elimination*: Assessors can prompt elimination by including more items in one list than the other. Learners may also be allowed to choose 'none' if an item has no relationship in the opposite list.

7.5.2   Other forms of matching-list questions

Matching-list questions are not confined to text items. Lists can be in the form of graphics. Assessors can express graphical relationships by dragging an object over another and dropping it there. Click-in-picture questions can be regarded as a fully fledged type of CIA.

7.6 CLICK-IN-PICTURE QUESTIONS

In click-in-picture questions, learners are required to choose an object or a particular point in a picture by pointing at it and clicking it with the mouse. In the example below, learners are taught to recognise the meaning of the colours of the South African flag.

The assessor should use click-in-picture tests instead of textual multiple-choice questions when it is more significant for learners to identify the position of something or the function of a particular area in a graphic.

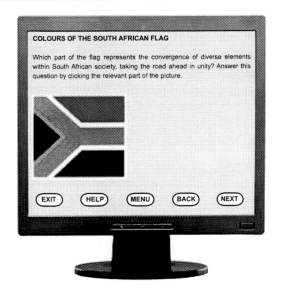

7.6.1 Formulating click-in-picture questions

When formulating click-in-picture questions, the assessor should:
- Explain clearly what learners are supposed to choose. Should they point at and click a specific object, icon, area or point?
- Ensure that targeted areas are visually distinguishable. The assessor should make them visually distinct graphics or areas with visible borders.
- Enlarge the target areas/objects so that they allow learners with average eye-hand coordination to select them without delay.
- Display the scene the way it would be in the real world. This can be done through games and simulations.
- Design graphics in a way that enables downloads to come quickly.

7.7 DRAG-AND-DROP QUESTIONS

The assessor should employ drag-and-drop questions in CIA to allow learners to move icons or images to specific areas on the screen. This will enable the assessor to test learners' competence to put items in a matching category or to organise the components of a system as a whole. In the example on the subject of leadership, learners are required to drag pictures (and the name) of the chief executive officers (CEO's) of some prominent South African organisations.

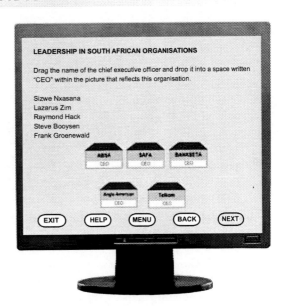

7.7.1 Formulating drag-and-drop questions

- Part of formulating drag-and-drop questions is devising understandable instructions for learners so that they know what they are expected to do. Prompt learners to drag the graphics to their specific slots.
- Indicate clearly pieces that learners can drag-and-drop. These pieces should be described in the instruction and in text.
- Identify the targeted slots clearly by making them visually distinct.

7.8 SIMULATION QUESTIONS

Assessors use simulation questions to let learners demonstrate their skills and knowledge interactively, but the use of this type of question was constrained mostly by poor bandwidth. In the example below, on the subject of air piloting, learners are required to demonstrate – by pointing at and clicking the relevant areas in the aircraft – how they would get it off the ground.

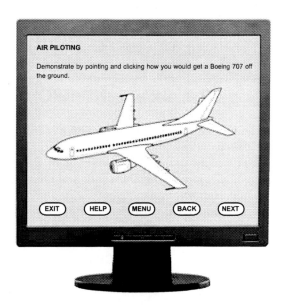

According to facilitators, a simulation enables learners to perform complex interactive tasks. If learners accomplish the required assignment by demonstrating through the simulator, they are then regarded as competent and can presumably carry out the real activity. It should be borne in mind, however, that simulations are generally expensive and difficult to design and develop. Despite that, experience shows that it is worth it. In fact, the benefits of using simulation questions eclipse the costs of design and development. Simulation questions meet the requirements of OBET exceptionally well.

7.8.1   Formulating simulation questions

Practitioners claimed that effective simulation questions should primarily require learners to demonstrate their skills and knowledge practically and interactively. When formulating the simulation questions, the assessor should:
- Ensure that the simulation is simplified. In CIA, simulation should be confined to testing knowledge and skills rather than teaching new knowledge and skills. On the other hand, the availability of choices should be moderate.
- State the use of simulation tests clearly. Learners should be provided with relevant instructions and information. These should include parameters, methods and features that can be used during the simulation assessment.

- Explain the simulation assessment prior to the actual assessment. Learners should know the potential and interactivity of simulation. The assessor should also familiarise learners with buttons, knobs and switches that they will use during the assessment.
- Inform learners about the defects of the simulation assessment. A simulation cannot be an exact copy of the thing it simulates. The assessor should reveal how the simulation affects learners' participation in the real context. In the military environment, for example, learners should be told how dangerous it can be to use explosives in the real world. The dangerous aspects of explosives, however, would be negated in the simulation context.
- Provide the scoring criteria. The method of awarding points should be spelled out. Learners will then know what they are expected to achieve. On the other hand, the criteria for awarding marks should allow the assessor to mark objectively.

7.9 FILL-IN-THE-BLANKS QUESTIONS

Fill-in-the-blanks questions in CIA require learners to put missing words into a sentence of text or in a missing piece of a table. Education, training and development managers confirmed that fill-in-blank questions have been in use for learning for many years. The example below, on the subject of remuneration, illustrates the use of CIA in the corporate training environment.

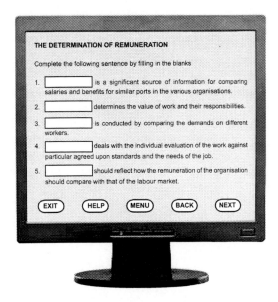

7.9.1 Formulating fill-in-the-blanks questions

Fill-in-the-blanks questions should be formulated in a simple and predictable manner, so that learners can focus on responding to the question. When formulating this type of question, the assessor should:

- Introduce the context to learners. The assessor should explain where the incomplete content comes from and its objectives.
- Provide learners with a list of possible alternatives.
- Ensure that the question is not ambiguous.

7.10 EXPLAINING THE COMPUTER-INTEGRATED ASSESSMENT PROGRAM

During interviews and observations, it was noted that in some instances the instructions were ambiguous or not understandable. If learners do not understand the assessment instructions, they may perform poorly, and the designers of the CIA should carry the blame. The instructions must be clear.

7.10.1 Providing instructions for computer-integrated assessment

Learners confirmed that it is difficult to undertake a CIA program if one does not understand the rules, regulations and restrictions before one starts the assessment. Ideally, instructions of the assessment program should:
- Indicate how a particular CIA program would be graded.
- Explain what the assessment program covers.
- Indicate when the assessment program can be done.
- Ensure that the length of the expected answer and the requirements of the questions match the marks allocated.
- Provide instructions regarding the accuracy of the responses. This includes guidelines for spelling and capitalisation.
- Inform the learner how many times he/she can retake the assessment.
- Inform learners about the resources that they can use during the course of the assessment program. These may include calculating devices, computer-integrated programs, books, Internet links, and other sources of information.
- Explain how the assessment will be graded. Assessors should indicate who will mark the assessment: the computer program, the assessor, other learners, or the candidate himself/herself. Learners should be told when they will be given the results of the assessment.
- Indicate whether the questions will be taken in sequence or randomly. The assessor should also explain whether moving to the next question leads to automated marking of the current question.
- Provide advice and guidelines on how one should respond to computer software and hardware failures.

The above list is very long for one assessment session. Assessors assert that the list of instructions should not be too long, as this may confuse and bore learners rather than be helpful. Instructions should be separated from the questions.

7.10.2 Principles of designing a computer-integrated assessment program

Only two of the organisations that participated in this study had documented principles regarding the design of CIA programs. These principles include:

- *Spelling out the rules*: The assessor should clearly spell out the rules/instructions to learners. This will prevent learners from making wrong assumptions.
- *Guarding against computer malfunctions*: Assessors and technicians who are involved in designing CIA should take steps to eradicate computer malfunctions during the assessment program. Observations confirmed that computer malfunctions can lead to fear, confusion and panic. Competent learners may then perform poorly.
- *Keeping the learner in control*: Learners should feel in control during the assessment program. Providing appropriate instructions will help. However, where pre-training in the use of computers has not been adequately provided, learners will not be in control during the program. This is supported by experience in the South African corporate training environment.
- *Phrasing questions precisely and clearly*: If the question is not clearly and unambiguously phrased, learners are vulnerable to wrong assumptions. A wrongly inserted word or punctuation mark may cause misunderstanding. The assessment program should therefore be presented to a subject expert for moderation and a language practitioner for editing. The assessor should pose one question at a time. In a corporate training environment, it is advisable to ask learners questions that are relevant. This enables learners to apply skills and knowledge in their working environment.
- *KISS – keep it simple for learners*: Assessors should avoid asking difficult, complicated and sophisticated questions that learners may struggle to interpret. This can be achieved by using simple sentences and common terms.
- *Shying away from obsolescence*: Assessors should guard against questions to which responses may change over time.
- *Treating absolute questions with care*: The problem with absolute questions and answers is that they tend to assume rigid categories.
- *Emphasising important words*: The assessor should emphasise essential or critical words in the question. Learners may wrongly interpret such words.
- *Making all choices plausible*: Multiple-choice questions should be formulated with care to offer plausible options to learners who may be tempted to guess. On the other hand, learners may remember the wrong answers rather than the correct ones. Options should also be more or less of the same length.
- *Challenging learners*: In the corporate training environment, questions tended to be too simple. Thus learners who did not have the required knowledge, skills, value and attitudes sometimes passed the assessment program easily.
- *Preventing one question from answering another*: Ensure that one question does not unintentionally provide the answer to another question.
- *Streamlining assessment questions*: Questions should be formulated simply and clearly so that learners do not waste time interpreting them and their badly worded optional answers. Choices should be kept fairly short.

7.10.3 Guidelines on providing meaningful feedback

There are several things that the assessor should consider when providing feedback to learners:

- Provide complete information. The assessor should avoid vagueness when providing the feedback. Give the correct answer. In some instances, explain why the learner's answer is incorrect.
- Be brief when explaining why a certain answer is the right one. Learners should be told why they are correct.
- Be gentle when remarking on the wrong answer. However, the error involved should be stated clearly.
- Avoid embarrassing, condemning or insulting the learner.
- Acknowledge correct responses. Learners should also be encouraged to try again.
- Provide a hint first in a situation where learners guess incorrectly. Give a hint and challenge the learner to attempt an answer again.

7.10.4 Online assessment methods

a) To assess knowledge and skills

Facilitators in this study believed that computers added considerable value to the way in which they assessed their learners: 'So online impacted on assessment rather than the other way round.' They used CIA to evaluate whether learners had acquired the necessary knowledge, skills, values and attitudes that the program was supposed to instil and offer. To demonstrate achievement, learners had to complete pre- and post-tests covering the relevant learning outcomes: 'The ability to conduct Kirkpatrick level 2 assessments wherever and whenever has reduced the classroom time spent on this and therefore raised the amount of actual learning time in these sessions.' The use of CIA to evaluate Kirkpatrick level 2 (knowledge and skills acquisition) is supported by authors such as Kruse and Keil (2000:84) and Vos (2000:235). CIA enables facilitators to assess knowledge and skills 'more frequently and thereby control the pace and quality of the learning'.

b) To administer assessment

The facilitators who were polled used computer-integrated tools to administer assessment: 'The learning management system tracks each learner scoring and assesses performance at various levels.' The learner performance reports were also automatically generated: 'Learning management system generates standard reports which make the administration of learning and assessment user friendly.'

c) To assess continuously

Facilitators preferred to have online assessment throughout the lesson: 'Pre-assessment, post-assessment – ongoing throughout the course.' However, assessment tends instructionally to come at the end of the process: 'We're excited about creating the content, we forget how to link the assessment to the content.' Most facilitators tried to have continual assessment. In fact, they stated that 'there should be pre-assessment in terms of recognition of prior learning'.

d) To assess the application of knowledge and skills

The facilitators said that they tried to assess the application of knowledge and skills: 'Assessment is limited in an online circumstance because you can only check knowledge.' Facilitators found it difficult to 'check application of that knowledge'. They also noted that learners usually performed well on post-tests, but the essence was whether they could retain, transfer and apply the new knowledge and skills to their work activities. Facilitators said that they were 'struggling with' assessing learners 'in the workplace, but logistics in terms of arranging that was just so much that [they] often didn't get around it'. Nonetheless, the availability of the learning management systems enabled them to 'make on-the-job assessments'. They said that it would be much easier 'to get the results from the person'. The need to assess the application of knowledge and skills is supported by Setaro (2003) and Kruse (2002a; 2002b).

7.11 FLEXIBLE ASSESSMENT STRATEGIES

Learners found that online learning enabled facilitators to employ flexible assessment strategies. Learners believed that the online assessment strategy was 'very good'. This could be attributed to the fact that there was continuous assessment 'because after learning such a chapter, you have to be tested'. The learners felt that the questions were very productive and thus they 'are not something you can memorise'. The nature of the questions was different and 'there are different tests, which are challenging'.

CHAPTER 8

FINAL WORD

CONTENTS

8.1 DISCUSSION OF THE FINDINGS

Most authors who are quoted in this book emphasise the use of a collaborative strategy in the online learning environment. However, the findings of the study indicate that a collaborative strategy has not been optimised. The few organisations that employ this strategy confine collaborative learning to asynchronous learning. In an ideal situation, learners should be given opportunities to learn collaboratively in asynchronous and synchronous situations. In reality, most South African organisations did not even consider employing this strategy, although their learning management systems have tools for it. CIE in the South African corporate environment is still in its infancy, so possibly practitioners are on a learning curve.

The findings indicate that learners were concerned about lack of competency among instructors in facilitating CIE. Learners suggested that facilitators should be given sufficient training, since their competence has an impact on learner proficiency and lack of competence results in the underutilisation of computer-integrated tools. Lack of sufficient training and support is therefore negatively affecting the use of a collaborative strategy. Various authors support this assertion. The findings also show that, as some experts have suggested, research should be undertaken to investigate the efficacy of collaborative learning in the corporate training environment. However, research conducted in schools and higher education institutions shows that collaborative learning adds considerable value to a learning environment, so one wonders whether the suggested research is really necessary.

Although the integration of CIE in the corporate training environment is still in its early stages, corporate South Africa is very advanced in certain areas of e-learning integration. Endeavours to integrate online learning with other applications such as electronic performance support systems, knowledge management applications and e-business initiatives are something new in the field of CIE and human resources development. South African organisations should therefore be commended for these endeavours. The integration of links is aimed at enhancing the skills and knowledge of employees with regard to their daily activities. The researcher has observed that the literature on CIE does not have much information on the issue of integrating e-learning with other systems in the organisation, possibly because the integration of systems is a new phenomenon in the field of human resources development. Although Morrison (2003) and Cronjé and Baker (1999) have written about the trend of integration, literature on this aspect is generally very limited.

The researcher found that some organisations had undertaken benchmarking trips to the US and Canada, where they interacted with practitioners and academics. It is puzzling to note that South African CIE practitioners in the corporate world do not interact with academics in South African higher education institutions.

The findings show that online learning practitioners were concerned about bandwidth constraints and understanding legislation that deals with content design and development. This could be attributed to a lack of interaction between stakeholders. Interaction between stakeholders will help to solve some of these problems.

8.2 CONCLUSION

The findings contained in this book would appear to indicate that organisations in South Africa are fairly competent in the integration of CIE for human resources development in the corporate training environment. In other words, CIE, training and development are adding value in the transformation of human resources. Managers and facilitators are employing various strategies, techniques and computer-integrated tools in their endeavours to integrate CIE in their training environments. Initiatives to integrate learning management systems with other applications such as electronic performance support systems and knowledge management applications are adding a lot of value to the success of online learning integration in the corporate training environment. This confirms that organisations have a broader view of the integration of online learning.

8.3 FINAL REMARKS

In seeking to articulate his feelings upon completing this book, the researcher is humbled by realising that what has been achieved is but a very small drop in the ocean of CIE and human resources development and that the following points apply:

- *CIE:* That organisations in South Africa are competent in integrating CIE for human resources development purposes holds great promise for future productivity.
- *The learning journey:* Although the book has been concluded, the scholastic and research journey has just started. What has been found and recommended in this study reinforces the conviction that much must still be done and researched, and that the final word will be a product of the collaborative efforts of more than one novice researcher.
- *Ideas:* Thoughts and ideas expressed and recommended in this study will make a difference only if they are tested and implemented.
- *This is our book:* An anonymous author once said: 'Certain authors, when they speak of their work say: "My book, my commentary, my history"; they would do better to say, "Our book, our commentary, our history"; since their writings generally contain more of other people's good things than of their own.' This researcher, along with the anonymous author, says: 'This is our research book.'

REFERENCES

Abell, M. and Foletta, G. Kentucky educators first to web with middle school and high school online learning. *Mathematics Teacher* 95(5)2002:396–397.

Adams, D. Wireless laptops in the classroom. *Communications of the ACM.* New York: ACM Press 49(9)(2006):25–27.

Adkins, S. S. The brave new world of learning. *Training and Development* 57(6)(2003):28–38.

Africare 2002. Closing the digital divide in South Africa. Available online: www.ifla.org/IV/ifla68/papers/038-134e.pdf (accessed May 2007).

African Management Programme 2003. Critical factors determining the success of managers. Pretoria: Unisa Centre for Business Management.

Aggarwal, A. K. and Bento, R. Web-based education. In Aggarwal, A. K. (ed). *Web-Based Learning And Teaching Technologies: Opportunities And Challenges.* London: Idea Group Publishing, 2000.

Alessi, S. M. and Trollip, S. R. *Computer-Based Instruction: Methods And Development.* Englewood Cliffs, NJ: Prentice Hall, 1991.

Alessi, S. M. and Trollip, S. R. *Multimedia for Learning: Methods and Development.* Boston, Mass: Allyn & Bacon, 2001.

Alexander, S. and Boud D. Learners still learn from experience when online. In Stephenson, J 2001: *Teaching And Learning Online: Pedagogies For New Technologies.* London: Kogan Page, 2001.

Allerton, H. E. All about E. *Training and Development* 56(10)(2002):6.

Andrian, C. M. *Distance Learning Technologies: Issues, Trends And Opportunities.* London: Idea Group, 2000.

Ankiewicz, P. The planning of technology education for South African schools. *International Journal of Technology and Design Education 5* 1995:245–254.

Anon. NEA working on criteria that will judge the quality of online learning. *Electronic Education Report* 8(11)(2001):3–4.

Anon. Benchmarks for virtual learning. *NEA Today* 19(7)(2001):21.

Anon. Tips for increasing e-learning completing rates. *Workforce* 80(10)(2001):56.

Arnone, M. Mixing and matching distance education software. *Chronicle of Higher Education* 48(37)(2002):33–34.

Arnone, M. Fathom adds training to distance education offerings. *Chronicle of Higher Education* 48(24)(2002):27.

Asmal, K. Address on 'Reversing the braindrain' by minister of education. Pretoria: Republic of South Africa, 1999.

Asmal, K. *Human Resources Development Strategy.* Cape Town: South African National Parliament, 2001.

Asmal, K. *Information society developments in education.* In *Proceedings of the International Conference on Technology and Education Africa.* Potchefstroom University for Christian Higher Education, 2002.

ASTD. *State Of The Training Industry Report.* Johannesburg: ASTD Global Network South Africa, 2003.

Avert. HIV and AIDS in South Africa. Available online: http://www.avert.org/aidssouthafrica, 2006 (accessed May 2007).

Bankseta. E-learning project final report. Banking Sector Education and Training Author-ity. University of the Western Cape: Department of Information Systems, 2003a.

Bankseta. *'E-learning 2003': Trip report – A business study program to Canada and the USA to experience best practices, innovation and leadership from e-learning practitioners.* Johan-nesburg: Bankseta, 2003b.

Barks-Ruggles, E., Fantana, T., McPherson, M., Whiteside, A. *The Economic Impact of HIV/ AIDS in Southern Africa.* Washington, DC: The Brookings Institution, 2001.

Barnum, C. and Paarmann, W. Bringing induction to the teacher: A blended learning model. *The Journal* 30(2)(2002):56–60.

Baxter, K. Online learning. *American Artist* 65(708)(2001):14–16.

BBC News. South African hit by 'brain-drain'. London: BBC, 2002.

Beffa-Negrini, P. A., Cohen, N. L. and Miller, B. Strategies to motivate students in online learning environments. *Journal of Nutrition Education and Behavior* 34(6)(2002):334–337.

Bhagwana, J. and Wall, K. 2006. Going with the franchising flow. *Business Day.* 4 July 2006.

Blaine, S. New group to tackle skills base upgrade. *Business Day*, 29 July 2006.

Bollinger, L. and Stover, J. *The Economic Impact of AIDS in South Africa.* Johannesburg: The Policy Project, 1999.

Bonk, C. J. and Cunningham, D. J. Searching for learner-centred, constructivist, and so-ciocultural components of collaborative educational learning tools. In Bonk, C. J. and King, K. S. (eds) 1998. *Electronic Collaborations: Learner Centred Technologies for Literacy, Apprenticeship And Discourse.* London: LEA, 1998.

Booysen, F., Van Rensburg, D., Bachmann, M., Engelbrecht, M., and Steyn, F. The socio-economic impact of HIV/AIDS on household in South Africa. *AIDS Bulletin.* Pretoria: South African Medical Research Council, 2002.

Botha, J. An analysis of the development of critical thinking during the presentation of a web-based course. MEd dissertation, Rand Afrikaans University, Johannesburg (un-published), 2000.

Boxer, K. M. and Johnson, B. How to build an online learning center. *Training and Develop-ment* 56(8)(2002):36–43.

Bridges, D. Back to the future: The higher education curriculum in the 21st century. *Cam-bridge Journal of Education* 30(1)(2000):37–56.

Brown, S. A. and Lahoud, H. A. An examination of innovative online lab technologies. *Proceedings of the 6th Conference on Information Technology Education*, 2005:65–70.

Bureau of Market Research. *The Projected Economic Impact of HIV/AIDS in South Africa, 2003 to 2015.* Pretoria: Unisa.

Burrows, T. Retaining trained staff. *IT Training Guide.* Rivonia: ITWeb Limited, 2001a.

Burrows, T. The case for e-learning. *IT Training Guide.* Rivonia: ITWeb Limited, 2001b.

Burrows, T. New buzzword – familiar strategy: Blended learning is simply evolved train-ing – not an epitaph for e-learning. *IT Training Guide.* Rivonia: ITWeb Limited, 2002.

Business Learning Institute. *Introduction to elearning.* Randburg: Elearning Institute, 2003.

Campbell, J., Sawert, B. and McPhee, L. Responding to the eArmy. *WebCT 2002* 4th Annual User Conference. Boston, Mass, 2002.

Canas, A. J. Terraforming cyberspace: Making the Web hospitable. *Proceedings of the 4th Annual Conference: 4–6 September 2002.* Bellville: University of Stellenbosch Business School, 2002.

Capper, J. The emerging market for online learning: Insights from the corporate sector. *European Journal of Education* 36(2)(2001):237–246.

Cara, S. Adult and community learning: What's in a name? *Adults Learning* 14(2), 2002.

Carlson, C. Virtual simulation tool on top. *E-Week* 20(31)(2003):32–33.

Carnevale, D. Assessment takes centre stage in online learning. *Chronicle of Higher Education* 47(31)(2001):43–45.

CCDD Quality Assurance Framework (Technikon SA). Florida: Technikon Southern Africa, 2003.

CETA. *Assessor, Advisor And Moderator: Recognition Of Prior Learning*. Johannesburg: Learning Performance Link, 2001.

Chambers, J. A. This time, we get to do it right. *Community College Week* 14(20)(2002):8–9.

Charp, S. Online learning. *The Journal Online* 29(8)(2002):8–9.

Chau, P. Y. and Hu, P. J. Examining a model of information technology acceptance by individual professionals: An exploratory study. *Journal of Management Information Systems* 18(4)(2002):191–229.

Chernobilsky, E., Nagarajan, A. and Hmelo-Silver, C. E. Problem-based learning online: multiple perspectives on collaborative knowledge construction. *Proceedings of the 2005 Conference on Computer Support for Collaborative Learning: Learning 2005: The Next 10 Years!*, 2005.

Christner, T. A classroom of one (book). *Library Journal* 128(1)(2003):130.

Cloete, E. and Miller, M. G. Structured educational design and modelling for a digital course. *South African Computer Journal* (28)(2002):60–66.

Cohen, R. *Brain Drain Migration*. Coventry: University of Warwick, 1997.

Collis, B. and Moonen, J. *Flexible Learning in a Digital World*. London: Kogan Page, 2001.

Conee, J., Shackelford, B., Boxer, K. M., Johnson, B. and Weaver, P. Executive summaries. *Training and Development*, 56(8)(2002):74–75.

Crane, B. E. Teaching with the Internet: Strategies and models for K-12 curricula. New York: Neal-Schuman, 2000.

Cronjé, G. J. De J. and Fourie, L. J. *African Management Program: Introduction to the Business World*. Pretoria: Unisa Press, 2003.

Cronjé, J. C. and Baker, S. J. B. Electronic performance support system: Appropriate technology for the development of middle management in developing countries. *South African Computer Journal* (28)(1999):42–53.

Crook, C. Still talking at the boundaries. *Journal of the Learning Sciences* 8(3/4)(1999):517–519.

Dagada, R. E-learning in corporate South Africa: Lessons learned. Paper delivered at the 'Magic in the Mix?' Seminar organised by the Elearning Institute, 1 October 2003.

Dagada, R. Educator competence in integrating computers for teaching and learning within the framework of the GautengOnline project. MEd dissertation, Rand Afrikaans University, Johannesburg, 2004.

Dagada, R. and Jakovljevic, M. 'Where have all the trainers gone?' E-learning strategies and tools in the corporate training environment. *Proceedings of The 2004 Annual Research Conference of The South African Institute of Computer Scientists and Information Technologists on IT Research in Developing Countries*. Pretoria: SAICSIT, 2004:194–203.

D'Amico, L. Networking CEO predicts e-learning wave. *InfoWorld* 21(49)(1999):4–65.

Denis, G. and Jouvelot, P. Motivation-driven educational game design: applying best practices to music education. *Proceedings of the 2005 ACM SIGCHI International Conference on Advances in Computer Entertainment Technology*, 2005.

De Klerk, F. W. Preparing for the new millennium. In *Proceedings of the International Conference on Technology and Education in Africa*. Potchefstroom University for Christian Higher Education, 2002.

De Lima, F. Web-based learning more cost effective. *Computing Canada* 25(27)(1999):29.

Den Biggelaar, J. C. M. Educating knowledge engineering professionals. In Liebowitz, J and Wilcox, L. C. (eds). *Knowledge Management and Its Integrative Elements*. New York: CRC, 1997.

Department of Education and Department of Communication (Republic of South Africa). *Strategy for Information and Communication Technology in Education*. Pretoria: Departments of Education and Communication, 2001.

Digital Partnership. The digital partnership facilitates innovation in affordable access to technology, training and the Internet for learning, enterprise and development. Available online: www.tve.org/ho/doc.cfm?aid=1608 - 22k (accessed: May 2007).

Din, H. W. H. Play to learn: exploring online educational games in museums. Material presented at the ACM SIGGRAPH 2006 Conference. New York, 2006.

Drysdale, E. Challenges in implementing e-learning. In Proceedings of the International Conference on Technology and Education Africa, Potchefstroom University for Christian Higher Education, 2002.

Elen, J. and Lowyck, J. Homogeneity in students' conceptions about the efficiency of instructional interventions: Origins and consequences for instructional design. *Journal of Structural and Intelligent Systems* 14(3)(2000):253–266.

Engelbrecht, E., Du Preez, C., Rheeder, R. and Van Wyk, M. Using an outcomes-based philosophy in the development of quality learning materials at Technikon SA: An action research report. In Proceedings of the 2nd National NADEOSA Conference, 2000.

Ensor, L. Skweyiya calls for basic income grant for poor: Manuel has warned against fostering culture of dependency. *Business Day*, 10 November 2006.

Ferguson, J. Government papers aim to bring copyright laws into digital age. *Billboard* 113(30)(2001):76.

Ferguson, R. B. Automating the back office. *E-Week* 20(31)(2003):16.

Ferguson, R. B. Buyers reach out. *E-Week* 20(31)(2003):31.

Fichter, D. Intranets and elearning: A perfect partnership. *Online* 26(1)(2002):68–71.

Fisch, S. M. Making educational computer games 'educational'. *Proceedings of the 2006 Conference on Interaction and Children*. New York: ACM Press, 2005.

Forsyth, T. *Teaching and Learning Materials and the Internet*. London: Kogan Page, 1998.

Friendlein, A. *Web Project Management: Delivering Successful Commercial Web Sites*. San Francisco, Calif: Morgan Kaufmann, 2001.

Fuller, A., Awyzio, G. and McFarlane, P. The benefits of increasing the integration of WebCT. *Proceedings of the 4th WebCT User Conference*. Boston, Mass, 2002.

Gagné, R. M. Computer-based instructional guidance. In Spector, J. M., Polson, M. C. and Muraida, D. J. (eds). *Automating Instructional Design: Concepts and Issues*. Englewood: Educational Technology Publishers, 1993.

Gagné, R. M., Briggs, L. J. and Wager, W. W. *Principles of Instructional Design*. London: Harcourt Brace, 1992.

Galvao, J. R. Martins, P. G. and Gomes, M. R. Modelling reality with simulation games for a cooperative learning. *Proceedings of the 32nd Conference on Winter Simulation: Driving Innovation*, 2003.

Garten, J. E. Intellectual property: New answers to new problems. *Business Week*, 3726 2001:28.

Gauteng Department of Education (GDE). *Draft Assessment Policy Document*. Johannesburg: Gauteng Department of Education, 1999.

Gauteng Department of Education (GDE). *National Assessment Policy As It Relates to OBE and the Implementation of Curriculum 2005 and Assessment In GET Grades.* (Circular 5/2000). Johannesburg: Gauteng Department of Education, 2000.

Gehring, J. Higher ed's online odyssey. *Education Week* 21(35)(2002):27–29.

Global Poverty Research Group. *Unemployment, Race and Poverty in South Africa.* Oxford: Oxford University Press, 2006.

Gokhale, A.A. Collaborative learning enhances critical thinking. Digital Libraries and Archive, 1995.

Gold, M. Entreprise e-learning. *Learning Circuits* – ASTD's online magazine all about e-learning, 2003. Available online: http://www.learningcircuits.org/2003/apr2003/gold.html (accessed May 2007).

Goral, T. Going the distance. *Curriculum Administrator* 37(4)(2001):52–55.

Grant, R. M. Shifts in the world economy: The drivers of knowledge management. In Despres, C. and Chawel, D. (eds) 2000. *Knowledge Horizons: The Present and Promise of Knowledge Management.* Boston: Butterworth-Heinemann, 2000.

Halverson, R., Shaffer, D., Squire, K. and Steinkuehler, C. Theorizing games in/and education. *Proceedings of the 7th International Conference on Learning Sciences.* International Society of Learning Sciences, 2006:1048–1052.

Hamilton-Pennell, C. Getting ahead by getting online. *Library Journal* 127(19)(2002):32–35.

Hannafin, M. J. and Peck, K. L. *The Design, Development and Evaluation of Instructional Software.* London: Collier McMillan, 1988.

Hargis, J. Can students learn science using the Internet? *Journal of Research on Computing in Education* 33(4)(2001):475–488.

Harris, P. Wake-up call. *Training and Development* 56(9)(2002):24–30.

Hill, B. Entered apprentice: a Luddite 'factory' for undergraduate learning in computer animation. *ACM SIGGRAPH 2006 Educators program.* New York: ACM Press, 2006.

Hofmann, J. Blended learning case study. *Learning Circuits,* 2001. Available online: http://www.learningcircuits.org/2001/apr2001/hofmann.html (accessed May 2007).

Hofmann, J. Peer-to-peer: The next hot trend in e-learning? *Learning Circuits,* 2002, and online: http://www.learningcircuits.org/2002/jan2002/hofmann.html (accessed May 2007).

Hofmann, J. Motivating online learners. *Learning Circuits,* 2003. Available online: http://www.learningcircuits.org/2003/aug2003/hofmann.html (accessed May 2007).

Hofmann, J. Peer-to-peer: Creating collaboration. *Learning Circuits,* 2003. Available online: http://www.learningcircuits.org/2003/sep2003/hofmann.html (accessed May 2007).

Honey, P. Racing towards fusion. *Financial Mail.* 1999:104–107.

Hoole, D. and Hoole, S. R. Web-based teaching: Infrastructure issues in the third world. In Aggarwal, G. (ed). *Web-Based Learning and Teaching Technologies: Opportunities and Challenges.* London: Idea Group, 2000.

Horibe, F. *Managing Knowledge Workers: New Skills and Attitudes to Unlock the Intellectual Capital in Your Organisation.* New York: John Wiley, 1999.

Horton, W. *Designing Web-Based Training: How to Teach Anyone Anything Anywhere Anytime.* New York: Wiley, 2000.

Hotler, D. Systems integration broaden their horizons. *IE Industrial Engineer* 35(6)(2003):35–38.

Independent Examination Board. Course for assessors. Johannesburg: Independent Examination Board Assessment Education and Training Department, 2005.

Jackson, D. New skills drive demands that Seta deliver: the government's accelerated initiatives will put existing training bodies. *Sunday Times* 2006.

Jain, S. and McLean, C. Simulation for emergency response: a framework for modelling and simulation for emergency response. *Proceedings of the 35th Conference on Winter Simulation: Driving Innovation*, 2003.

James, G. W. Take the ID road to success. *Training and Development* 55(4)(2001):16–17.

Jefferson, S. Two approaches to web-based learning. *InfoWorld* 22(30)(2000):54–55.

Jenkins, S. An LMS helps meet HIP AA compliance. *Training and Development* 57(3)(2003):74–75.

Johnson, D. F. Developing effective WebCT administration practices at the University of Florida. *Proceedings of the WebCT 4th Annual User Conference*. Boston, Mass, 2002.

Johnson, R. Thinking global, acting local: Gauteng finds viable education solutions. *Convergence: GautengOnline Special Edition*. Parkwood: Connexity, 2002.

Johnson, R. B. Examining the validity structure of quantitative research. *Education* 118(2)(1997):282–293.

Johnson, S. D. Learning technological concepts and developing intellectual skills. *International Journal of Technology and Design Education* 7(1–20)(1997):161–180.

Kazmer, M. M. How technology affects students' departures from online learning communities. *ACM SIGGROUP Bulletin*. New York: ACM, 25(1)(2005):7–11.

Kazmer, M. M. and Haythornthwaite, C. Multiple perspectives on online learning. *ACM SIGGROUP Bulletin*. New York: ACM Press, 25(1)(2005):25–30.

Ke, F. Classroom goal structures for educational math game application. *Proceedings of the 7th International Conference on Learning Sciences*, 2006.

Kemery, E. R. Developing online collaboration. In Aggarwal, A. (ed) 2000. *Web-Based Learning and Teaching Technologies: Opportunities And Challenges*. London: Idea Group, 2000.

Kern, P. In EU, digital rights holders need protection. *Billboard* 113(4)(2001):4–5.

King Committee on Corporate Governance. *Executive Summary of the King II Report on Corporate Governance for South Africa*. Parktown: Institute of Directors in Southern Africa, 2000.

Kingdom, G. and Knight, J. *Unemployment in South Africa, 1995–2003: Causes, Problems And Policies*. University of Oxford, 2005.

Klobas, J. and Renzi, R. Selecting software and services for Web-based teaching and learning. In Aggarwal, A. (ed). *Web-Based Learning and Teaching Technologies: Opportunities And Challenges*. London: Idea Group, 2000.

Kruse, K. and Keil, L. *Technology-Based Training: The Art and Science of Design, Development, and Delivery*. San Francisco: Jossey-Bass/Pfeiffer, 2000.

Kwinn, A. The introverted trainer. *Training and Development* 55(9)(2001):22–24.

Lautenbach, G. V. Learner experiences of web-based learning: A university case study. MEd dissertation, Rand Afrikaans University, Johannesburg, 2000.

Lautenbach, G. and Van der Westhuizen, D. Professional development of the online instructor in higher education: A program for Web-based higher education. *Proceedings of the 4th Annual Conference: 4–6 September 2002*. Bellville: University of Stellenbosch Business School, 2002.

Lee, J., Luchini, K., Michael, B., Norris, C. and Soloway, E. More than just fun games: assessing the value of educational video games in the classroom. *Conference on Human Factors in Computing Systems*. New York: ACM Press, 2004.

Le Roux, C. Butchers or builders: Approaches to assessment in distance learning. *South African National Defence Force Bulletin for Educational Technology – Proceedings of the 2004 Department of Defence Education, Training and Development Conference*, 2004.

Le Roux, M. Skills shortage in construction sector a threat to World Cup. *Business Day*, 29 July 2006.

Leyell, T. S. The effectiveness of technology enhanced learning. *Proceedings of the WebCT 2002 4th Annual User Conference*. Boston, Mass, 2002.

Lindroth, L. Blue ribbon technology. *Teaching Pre K-8* 33(3)(2002):22–25.

Littleton, K. and Light, P. Introduction – Getting IT together. In Littleton, K. and Light, P. (eds). *Learning with Computers: Analysing Productive Interaction*. London: Routledge, 1998.

Ludlow, B. L., Foshay, J. D., Brannan, S. A., Duff, M. C. and Dennison, K. E. Updating knowledge and skills of practitioners in rural areas: A web-based model. *Rural Special Education Quarterly* 21(2)(2002):33–44.

Lynch, P. J. and Horton, S. *Web Style Guide: Basic Design Principles For Creating Web Sites*. New Haven: Gale University Press, 1999.

Madiope, M. and Dagada, R. Computer-integrated African Languages Program at the University of South Africa. E-Learn 2004 – World Conference on E-Learning in Corporate, Government, Healthcare, and Higher Education. Washington DC, 1–5 November 2004. Volume 6 (ISBN 1-880094-54-1)(2004):1734–1740.

Mail and Guardian. All we do is fight for a turf. *Mail and Guardian* 2003.

Mamaila, K. and Green, J. Intelligence agents to aid firms fight foreign spies. *The Star,* 6 November 2002.

Mamudi, S. WIPO calls for global action on domain names. *Managing Intellectual Property* 2001:113.

Marconi Communications. *ICT Education Project*. Johannesburg: Marconi Communications, 2002.

Mayberry, E. How to build a business case for e-learning. *Learning Circuits*, 2001. Available online: http://www.learningcircuits.org/2001/jul2001/mayberry.html (accessed April 2007).

Mbeki, T. M. *Address of the President of South Africa at the Conference of the Association of African Universities*. Cape Town: University of Cape Town, 2003.

Mbeki, T. M. *State of the Nation Address of the President of South Africa, Thabo Mbeki: Joint Sitting of Parliament*. Pretoria: Presidency, 2006.

McCormack, C. and Jones, D. *Building A Web-Based Education System*. New York: Wiley Computer Publishing, 1998.

McDermott, R. Subject matter expert. *Knowledge Management Review* 40(20)(2001):5.

McLean, D. *Everything You Need To Know About Computer Assisted Assessment*. Cape Town: The Institute of People Development, 2002.

McLester, S. Virtual learning takes a front row seat. *Technology and Learning* 22(8)(2002):24–31.

McLester, S. Virtual instruction a trade-off for teachers. *Technology and Learning* 22(8) 2002b: 4–7.

Metzler, J. The hidden value of strategic partnerships. *Internet Week* (75)(1999):27–29.

Miller, E. Protecting information. *Computer Aided Engineering* 20(9)(2001):59.

Minkel, W. Web of deceit. *School Library Journal* 48(4)(2002):50–53.

Mioduser, D., Nachmias, R., Lahav, O and Oren, A. Web-based learning environments: Current pedagogical and technological state. *Journal of Research on Computing in Education* 33(1)(2000):55–76.

Mlambo-Ngcuka, P. Address delivered by Deputy President Phumzile Mlambo-Ngcuka, Department of Home Affairs Conference on Foreign Offices Operations, Birchwood Hotel Conference Centre. Pretoria: Presidency, 2006.

Molnar, A. R. Computers in education: A brief history. *THE Journal Online*, 24(11)(1997):63–68.

Mona, V. It's less a luxury that you'd think. *City Press*, 18 June 1999.

Moore, C. E-learning gets boost from IBM. *InfoWorld* 23(35)(2001):34.

Moore, C. Web services tap e-learning. *InfoWorld* 25(7)(2003):22.

Morphew, V. N. Web-based learning and instruction: A constructive approach. In Lau, L. (ed). *Distance Learning Technologies: Issues, Trends And Opportunities*, Hershey, Pa: Idea Group, 2000.

Morrison, D. *E-Learning Strategies: How To Get Implementation And Delivery Right First Time*. Chichester: Wiley, 2003.

Murdoch, A. Right to copy? *World Link* 2001:33–34.

Musslewhite, C. Simulation classification system. *Learning Circuits* – ASTD's online magazine all about e-learning, 2003. Available online: http://www.learningcircuits.org/2003/aug2003/musselwhite.html (accessed May 2007).

Ngubane, B. Keynote address by the Minister of Arts, Culture, Science and Technology, Faculty of Medicine Graduation Ceremony of the Medical University of South Africa, Pretoria, South Africa.

Ntuli, Z. Asgisa boasts of 2500 seat call centre. *The Skills Portal*. Available online: www.skillsportal.co.za (accessed on May 2007).

Nxasana, S. Crossing the digital divide: One bridge at a time. *Proceedings of the International Conference on Technology and Education Africa*. Potchefstroom University for Christian Higher Education, 2002.

Oakes, K. and Rengarajan, R. E-learning. *Training and Development* (56(11)(2002):58–60.

O'Connell, B. A poor grade for e-learning. *Workforce* 81(7)(2002):15.

O'Neal, K.M. Why I am failing my junior officers. *Proceedings of the United States Naval Institute* 129(7)(2003):40–42.

Pack, T. and Page, L. Creating community. *Information Today* 20(1)(2003):27–28.

Palloff, R. M. and Pratt, K. *Lessons from the Cyberspace Classroom: The Realities of Online Teaching*. San Francisco: Jossey-Bass, 2001.

Panitz, T. A definition of collaborative vs co-operative learning, 1996. Available online: http:www.lgu.ac.uk/deliberations/colab.learning/panitz2.html (accessed on 17 July 2002).

Perry, M. Ensuring security the hard way. *Electronic Engineering Times* 1143(2001):108–109.

Persaud, A. The knowledge gap. *Foreign Affairs* 80(2)(2001):107–118.

Pethokoukis, J. M. E-learn and earn. *U.S. News and World Report*, 132(22)(2002):36.

Phillips, R. Educational considerations. In Phillips, R. (ed). 1997: *The developer's handbook*. London: Kogan Page, 1997:18–35.

Pile, J. Small business shrugs it off. *Financial Mail*, 23 April 2004.

Place, J. L., Stephens, T. and Cummingh, P. O. Delivering clinical-based training in a public health setting. In Schreiber, D. A. and Berge, Z. L. (eds). *Distance Training: How Innovative Organisations Are Using Technology To Maximise Learning And Meet Business Objectives*. San Francisco: Jossey-Bass, 1999.

Policy Co-ordination and Advisory Services. *Macro-Social Report: A Discussion Document On Macro-Social Trends In South Africa*. Pretoria: Presidency, 2006.

Poole, B. J. *Education for an Information Age: Teaching in the Computerised Classroom*. New York: WCB McGraw-Hill, 1998.

Porter, L. R. *Creating the Virtual Classroom: Distance Learning with the Internet*. New York: John Wiley, 1997.

Quilling, R. D. Erwin, G. J. and Petkova, O. Active learning: Issues and challenges for information systems and technology. *South African Computer Journal* (24)(1999):5–14.

Richards, R. WebCT: Chronicles of training. *Proceedings of the WebCT 2002 4th Annual User Conference*. Boston, Mass, 2002.

Rolland, N. and Chauvel, D. Knowledge transfer in strategic alliances. In Despres, C. and Chauvel, D. (eds). *Knowledge Horizons: The Present and Promise of Knowledge Management*. Boston: Butterworth-Heinemann, 2000.

Saunders, P. and Werner, K. Finding the right blend for effective learning, 2002. Available online: lttf.ieee.org/learn_tech/issues/april2002/index.html (accessed May 2007).

Savadis, A. and Stephanidis, C. Developing inclusive e-learning and e-entertainment to effectively accommodate learning difficulties. *ACM SIGACCESS Accessibility and Computing*. New York: ACM Press 83(2005):42–54.

Schneier, B. *Secrets and Lies: Digital Security in a Networked World*. New York: Wiley Computer Publishing, 2000.

Sein, M. K. and Simonsen, M. Effective training: applying framework to practice. *Proceeding of the 2006 ACM SIGMS CPR Conference on Computer Personnel Research: Forty-Four Years of Computer Personnel Research: Achievements, Challenges and the Future*. New York: ACM, 2006.

Setaro, J. Many happy returns: Calculating e-learning ROI, 2001. Available online: http://www.learningcircuits.org/2001/jun2001/elearn.html (accessed May 2007).

Shezi, A. Talk shops will not flay the jobs dragon. *Business Day,* 29 July 2006.

Shukla, R. and Koh, D. Transition from online support to online course: blending with ICT. *Proceedings of the Winter International Symposium on Information and Communication Technologies*. Trinity College Dublin, 2004.

Sinclair, K. The benefits of online learning. *Engineered Systems* 18(6)(2001):32.

Slabbert, J. A. and Fresen, J. W. Online communication problems, solutions and opportunities in e-learning. *Proceedings of the 4th Annual Conference: 4–6 September 2002*. Bellville: University of Stellenbosch Business School, 2002.

Smith, J. M. Blended learning: An old friend gets a new name, 2001. *Executive Online*. Available online: http://www.gwsae.org/Executiveupdate/2001/March/blended.htm (accessed May 2007).

South Africa. Accelerated and Shared Growth Initiative for South Africa. Pretoria: The Presidency, 2006.

South Africa. Constitution of the Republic of South Africa. Pretoria. South African Government Printer, 1996. Available online: www.info.gov.za/documents/constitution/index.htm.

South Africa. *Assessment Policy in the General Education and Training Band Grades R to 9 and ABET*. Pretoria: National Department of Education, 1998.

South Africa. Department of Communication. Electronic Communications and Transactions Act, 2002. Available online: www.info.gov.za/gazette/acts/2002/a25-02.pdf (accessed May 2007).

South Africa. Department of Communication. Government to address Internet inequalities and governance. *The Star*, Business Report, 17 December 2004.

South Africa. Department of Education. South African Qualifications Authority Act, 1995. Available online: www.acts.co.za/ed_saqa/index.htm (accessed May 2007).

South Africa. Department of Labour. Employment Equity Act, 1998a. Available online: www.workinfo.com/Free/Sub_for_legres/data/*equity*/Act551998.htm (accessed May 2007).

South Africa. Department of Labour. Skills Development Act, 1998b. Available online: http://www.labour.gov.za/docs/legislation/skills/act98-097.html (accessed May 2007).

South Africa. Department of Labour. Skills Development Levies Act, 1999. Available online: www.info.gov.za/gazette/acts/1999/a9-99.pdf (accessed May 2007).

South African Broadcasting Corporation (SABC). *Mbeki to study data on white poverty*, 2004. Available online: www.theherald.co.za/herald/2004/10/22/default.htm. (accessed May 2007).

South Africa. Labour Department will take action on Setas that do not perform. Statement released by Department of Labour on 19 March 2003. Pretoria: Department of Labour.

South Africa. Minister calls on SETAS to spend R2,8 billion surplus.

Statement released by Department of Labour on 10 July 2003.

St Clair, J. Tennessee regents online – Learning degrees program. *Proceedings of the WebCT 2002 4th Annual User Conference, 2002*. Boston, Mass, 2002.

Stephens, P. J. Teaching physiology in a virtual world. *Proceedings of the 4th Annual User Conference, 2002*. Boston, Mass, 2002.

Swanepoel, B.J., Erasmus, B.J., Kirsten, M. and Holtzhausen, M. *Labour Relations Management: A Macroperspective*. Pretoria: Unisa Centre for Business Management, 2003.

Tang, S. Jipsa training program launched at Old Mutual Business School. *The skills portal*. Available online: www. Skillsportal.co.za (accessed 20 September 2006).

Taylor, S. S. Education online: Off course or on track? *Community College Week* 14(20) (1999):10–12.

Technikon South Africa and South African Institute for Distance Education. 2002. *Evaluation Capacity – Building Project: Report Part 3 – Evaluation Instruments*. Technikon SA, 2002.

Tetiwat, O. and Igbaria, M. Opportunities in Web-based teaching: The future of education. In Aggarwal, A. (ed) 2000: *Web-Based Learning And Teaching Technologies: Opportunities And Challenges*. London: Idea Group 2000.

TFPL. *Skills for Knowledge Management: Building a Knowledge Economy*. New York: TFPL, 1999.

Tiffin, J. and Rajasingham, L. *In Search Of The Virtual Class: Education In An Information Society*. London: Routledge, 1995.

Toma, J. How getting close to your subjects makes qualitative data better. *Theory into Practice* 39(3)(2000):177–185.

Troha, F. J. Bulletproof instructional design: A model for blended learning. *USDLA Journal* 16(5)(2002).

Trotter, A. Calif's online-learning potential evaluated. *Education Week* 22(8)(2002):11–12.

Tsedu, M., Sikhakhane, J. and Jeffreys, H. The Thabo Mbeki interview. *City Press*, 11 April 2004.

Tshabalala-Msimang, M. Speech by minister of health during the signing of the Memorandum of Understanding between South Africa and UK. Pretoria: Republic of South Africa, 2003.

Tulleken, L. Government to address Internet inequalities and governance – minister. *The Star: Business Report,* 17 December 2004.

Tucker, B. *Handbook of Technology-Based Training.* Aldershot, Hampshire: Gower, 1997.

University of South Africa. *Assessment Policy: Draft 1.* Pretoria: Unisa, 2004.

Valiathan, P. Blended learning models. *Learning Circuits,* 2002. Available online: http://www.learningcircuits.org/2002/aug2002/valiathan.html (accessed May 2007).

Van der Spek. R. and Spijkervet, A. Knowledge management: Dealing with knowledge. In Liebowitz, J. and Wilcox, L. C. (eds). *Knowledge Management and its Integrative Elements.* New York: CRC 1997.

Van der Vyver, J. From rote learning to meaningful learning. *Education Practice* (5) (2000):39–43.

Van der Westhuizen, D. Teaching information technology in education using online education. DEd dissertation, Rand Afrikaans University, Johannesburg, 1999.

Van der Westhuizen, D. and Krige, H. Ending the divide between online learning and classroom instruction using a blended learning approach, 2002. Available online: http://general.rau.ac.za/cur/vdwesthuizen/published_conf_proc.htm (accessed May 2007).

Van der Westhuizen, D., Stoltenkamp, J. and Lautenbach, G. Help us! We want to 'e-teach': University lecturers support needs for facilitating e-learning. *Proceedings of the 4th Annual Conference: 4–6 September 2002.* Bellville: University of Stellenbosch Business School, 2002.

Verton, D. US talks cyber-security will be key with lawmakers. *Computerworld* (2002):35(38).

Von Broembsen, M. Time to get real about what survivalist enterprises need. *Business Day,* 2005.

Vos, H. How to assess for improvement of learning. *European Journal of Engineering Education* 25(3)(2000):227–234.

Wa Kivilu, M. Assessment of and instruction for higher order thinking skills. *Education Practice* (5)(2000):44–51.

Walker, S. The value of building skills online technology: Online training costs and evaluation at the Texas Natural Resource Conservation Commission. In Schreiber, D. A. and Berge, Z. L. (eds) 1999: *Distance Training: How Innovative Organisations Are Using Technology To Maximise Learning And Meet Business Objectives.* San Francisco: Jossey-Bass, 1999.

Warbington, R. The advantages of online learning. *Woman in Business* 53(6)(2001):23.

Weller, M. *Delivering Learning On The Net: The Why, What And How Of Online Education.* London: Kogan Page, 2002.

Wessels, J.S. Criteria for assessing learning materials for distance education. *South African Journal of Higher Education* 15(1)(2001):217–224.

Westerman, A. The relation between corporate training and development expenditures and the use of temporary employees. *Ethics and Behavior* 11(1)(2001):67–86.

Western Cape Education Department. The Khanya technology in education project, 2002. Online. Available: www.capegateway.gov.za/eng/pubs/speeches/2005/apr/102769 (Accessed May 2007).

Wheatly, G. H. Constructivist perspectives on science and mathematics learning. *Science Education,* 75(1)(1991):12–19.

Wiig, K. M. Knowledge management foundation – thinking about thinking: How people and organisations create, represent, and use knowledge. *Arlington: Scheme Press,* 1993.

Wiig, K. M. Knowledge management: An emerging discipline rooted in long history. In Despres, C. and Chavvel, D. (eds). *Knowledge Horizons: The Present And Promise Of Knowledge Management.* Boston: Butterworth-Heinemann, 2000.

Wilcox, L. C. Knowledge-based systems as an integrating process. In Liebowitz, J. and Wilcox, L. C. (eds). *Knowledge Management and Its Integrative Elements.* New York: CRC 1997.

Wild, M. Editorial: Accommodating issues of culture and diversity in the application of new technologies. *British Journal of Education Technology* 30(3)(1999):195–200.

Willis, M. and Kelly, D. All change, please. *Community Care* 1448(2002):38–39.

Wilson, D. H. Cast an eye on our own background. *Business Day,* 2006.

Woit, D. and Mason, D. Enhancing student learning through on-line quizzes. *Proceeding of the thirty-first SIGCSE technical symposium on computer science education.* New York: ACM Press, 2000.

Wolf, A. Circuit city, Marta pursue online sales training programs. *Twice: This Week in Consumer Electronics* 17(8)(2002):13.

Wolfe, C.R. Learning and teaching on the World Wide Web. In Wolfe, C. R. (ed). *Learning and Teaching on the World Wide Web.* New York: Academic Press, 2001:1–22.

Young, S. and McSporran, M. Facilitating successful online computing courses while minimising extra tutor workload. *Proceedings of the sixth conference on Australasian computing education.* Darlinghurst: Australian Computer Society, 2004.

Zafeiriou, G., Nunes, J. M. and Ford, N. Using students' perceptions of participating in collaborative learning activities in the design of online learning environments. *Education for Information* 19(2)(2001):83–107.

Zake, S. Graduates unsuited for real world. *Business Day.* 25 May 2006.

Zake, S. Nafcoc sets up business service centres. *Business Day.* 9 June 2006.

Zheng, D. and Young, M. Comparing instructional methods for teaching technology in education to preservice teachers using logistic regression. *Proceedings of the 7th International Conference on Learning Sciences.* International Society of the Learning Sciences, 2006.